D1146866

WILD FLOWERS

OF BRITAIN AND IRELAND

Rae Spencer-Jones learned to love wild flowers and her natural surroundings as a child from the many walks she enjoyed with her mother in the Buckinghamshire countryside. Rae studied horticulture at Writtle College before working for garden photographer Marcus Harpur and his father Jerry. After a period on the editorial team of *Gardens Illustrated*, she was a freelance horticultural journalist writing for, among others, the *Daily Telegraph, The Times, BBC Gardeners' World* and *Gardens Illustrated*. Rae is currently the book publisher for the Royal Horticultural Society.

Sarah Cuttle is one of the UK's most inspiring garden photographers. Her work is regularly published in national magazines, including *BBC Gardeners' World* and RHS *The Garden*. She is commissioned by the RHS to cover their annual shows, and has recently been accepted as associate photographer for RBG Kew. Sarah says that she has always been intrigued by the aesthetics of wild flowers – how features and colours can be used in the smallest of proportions – and also by how incredibly hardy they are, with an enviable way of adapting to their environment.

GREENWICH LIBRARIES

3 8028 02375397 0

WILD FLOWERS

OF BRITAIN AND IRELAND

RAE SPENCER-JONES AND SARAH CUTTLE

BOTANICAL CONSULTANT AND ADDITIONAL TEXT: BARRY TEBBS

Kyle Books

PREVIOUS PAGE, LEFT 'The heavens upbreaking through the earth', the English bluebell (*Hyacinthoides non-scripta*) is the quintessential wild flower.
PREVIOUS PAGE, RIGHT Herb paris (*Paris quadrifolia*), wild flower of woodland and shady places.
RIGHT A mass of ox-eye daisies, corncockles and cornflowers.

An Hachette UK Company
www.hachette.co.uk

This paperback edition published in Great Britain in 2018 by Kyle Books, an imprint of Kyle Cathie Ltd
Carmelite House, 50 Victoria Embankment
London EC4Y 0DZ
www.kylebooks.co.uk

First published in hardback in 2005
10 9 8 7 6 5 4 3 2 1

ISBN: 978 0 85783 474 4

All rights reserved. No reproduction, copy or transmission of this publication may be made without written permission. No paragraph of this publication may be reproduced, copied or transmitted save with written permission or in accordance with the provision of the Copyright Act 1956 (as amended). Any person who does any unauthorised act in relation to this publication may be liable to criminal prosecution and civil claims for damages.

Rae Spencer-Jones and Sarah Cuttle are hereby identified as the authors of this work in accordance with Section 77 of the Copyright, Designs and Patents Act 1988.

Text © 2005 Rae Spencer-Jones
Photography © 2005 Sarah Cuttle, see also p256
Illustrations © 2005 Lizzie Harper
Book design © 2005 Kyle Cathie Limited

Senior editor	Muna Reyal
Designer	Fran Rawlinson
Photographer	Sarah Cuttle, see also p256
Copy editor	Michael Wright
Editorial assistant	Cecilia Desmond
Illustrations	Lizzie Harper
Production	Sha Huxtable and Alice Holloway

A Cataloguing In Publication record for this title is available from the British Library.

Printed and bound in China by C&C Offset Printing Co., Ltd.

GREENWICH LIBRARIES

ELT

3 8028 02375397 0

Askews & Holts	09-Oct-2018
582.13	£20.00
5850943	

Contents

Introduction

For those of us fortunate enough to have grown up in the countryside or with a garden, wild flowers are inevitably woven into the fabric of our childhood. How many of us have memories of painstakingly constructing a daisy chain, holding a buttercup in hot sweaty hands beneath our chin to prove beyond doubt that we adore butter, or vigorously sucking on the flowers of a deadnettle to squeeze out the faintest trace of sweet nectar?

Wild flowers are the essence of the natural world, whether an azure sea of bluebells that stretches as far as the eye can see or a single rare and exquisite orchid that demands close inspection on hands and knees to admire its intricate detailing. Apparently so fragile, these plucky plants should never be underestimated. Wild flowers are brilliantly equipped to survive the most hostile of natural environments, from rock crevices to salt marshes. Their ability to attract and accommodate a whole host of pollinating insects is a miracle of engineering and, whether distributing their seed via minuscule feathery parachutes or gluttonous ants, each plant has evolved its own remarkable device to guarantee its reason for being – the ability to compete effectively and efficiently for survival.

In the face of all this genius it is unsurprising that wild flowers have been fêted by and invoked the admiration of botanists, herbalists, artists and poets for thousands of years. As a result, myriad wild-flower field guides, elaborate herbals and a wealth of literature and art have been produced in their honour.

Our aim for this book was first and foremost to create a friendly photographic guide for identifying wild flowers by colour. However, as the concept for the book evolved, we found ourselves not only amazed by the methods these plants employ to survive and flourish in their natural habitats, but captivated by the folklore surrounding wild flowers and inspired by the powerful influence that they have had on the foods we eat and the medicines we have relied on for centuries.

Yet sadly in the 21st century these precious plants are facing their greatest challenge: to survive the loss of their natural habitats through intensive farming methods, to compete against invasive foreign plant introductions and to combat indiscriminate and wholesale collection from the wild. All these factors are man-made and some are irreversible.

To this end, the chapter devoted to the conservation of wild flowers looks in more detail at the issues surrounding the threats to wild-flower populations, the measures in place to protect them and some of the many organisations involved in their conservation. The short chapter on gardening with wild flowers provides an

The glowing yellow blooms of bog asphodel (*Narthecium ossifragum*) turn dark orange as they mature before dramatic orange fruits appear in the autumn.

LEFT Famously immortalised by Wordsworth, drifts of the dainty wild daffodil (*Narcissus pseudonarcissus*) are now surprisingly rare in Britain.

A magnet for pollinating bees and butterflies, the purple-flowering marsh thistle (*Cirsium palustre*) is a common wild flower of marshes and damp woodland.

overview on recreating their habitats in your own backyard – the single most important contribution that we, as plant lovers and gardeners, can make to the conservation and protection of species (other than refraining from picking them).

However, important as they are, it is vital to remember that wild flowers are part of a wider picture, which includes the insects, birds and other animals they attract and benefit from and not least the habitats where both flora and fauna should be allowed to thrive undisturbed. Once equipped with the ability to identify wild flowers, we can then begin to understand and interpret our surroundings, the associated plant life and creatures and most importantly appreciate and protect these increasingly fragile and endangered ecosystems. This has to be the ultimate purpose of this book.

Producing *Wild Flowers of Britain and Ireland* has been a remarkable experience. Both of us have been buoyed along by our own deep-seated affection for wild flowers – its origins in our own childhood memories – along with a growing passion and respect for the plants that we have photographed and researched, and the issues that surround them. Between us we have learned more about wild flowers than we could have ever imagined, yet we are still painfully aware that we have only just scratched the surface. Through our words and pictures we aim to share with you some of what we have learned and in doing so, inspire in you an appreciation for the natural environment that will remain with you for a lifetime.

Rae Spencer-Jones and Sarah Cuttle, 2005.

About this book

Choosing the plants

Approximately 2,500 native species of wild flowers have been recorded in Britain and Ireland. To represent all these plants is beyond the scope of this book. Instead we chose just over 600 native and introduced flowering perennials, biennials, annuals, climbers and some sub-shrubs that you are most likely to come across in the rural or urban environment. Plants not included in this book were rejected on the basis that they were too similar to species already included in the list, too rare or too insignificant. The representation of wild plant families is as comprehensive as possible, but this book does not include grasses, sedges, ferns, shrubs or trees.

Britain and Ireland

This book covers wild flowers both native and introduced in Great Britain and Ireland. They are also found in much of the rest of northern Europe, including France, Germany, Sweden, Norway, Finland, Denmark, Belgium, Holland, the Faeroe Islands and Iceland.

The photographs

The photography for this book is the result of a Herculean effort. Sarah spent two years driving thousands of miles to capture wild flowers in all weathers and every kind of habitat. It took a 300-mile round trip to find the elusive early spider orchid on the Dorset coastline; she rolled up her trousers and waded into a lake near Snowdonia to get a shot of water lobelia, and braved London's South Circular to photograph a drift of wild daffodils. Occasionally Sarah employed eccentric methods, including receiving a sample of Bermuda buttercup by post in a Tupperware box. She also spent many painstaking hours in a glasshouse at

British Wildflower Plants in Norfolk under the expert eye of Linda Laxton. Linda is just one of the extremely knowledgeable botanists who gave generously, freely and patiently of their help and advice to produce this book, along with Dr Tom Cope and our botanical adviser Barry Tebbs.

The choice to photograph the plants rather than to illustrate them was made to give a true representation of the plants rather than an idealised version; we hope that this makes it easier for you to identify them.

The images were taken both on location and under studio conditions. The location shots were all taken in the British Isles and aim to set the wild flowers in their natural context, giving them a sense of place to make them easier to identify. The studio

Water lobelia (*Lobelia dortmanna*) thrives in shallow acid water where it throws up elegant spires of exquisite pale lilac flowers.

images consist of plants grown under nursery conditions at British Wildflower Plants in Norfolk page 247).

The 600 wild flowers were photographed over a period of two years and have been placed chronologically, as far as possible, within each chapter. In hindsight this record shows that wild flowers bloom when we least expect them to: early, late or not at all, depending on the climatic vagaries of each season. In Britain, 2003 presented us with a typically wet British summer until the end of July, when the rain was replaced by a searing drought causing the wild flowers to simply disappear halfway through the season. In 2004 the weather was wetter and warmer overall and as a consequence many plants flowered earlier, while later in the season a greater number of wild flowers were still in evidence than the previous year.

Colour
The wild flowers in this book have largely been arranged according to the colour of their flowers. Conventional wild-flower field guides are ordered by plant family, but as this book is not attempting to replicate a botanical field guide in the traditional sense, we have chosen to focus on the most obvious aspect of a plant – the colour of its flowers. In so doing we hope that we have made plant identification easier, particularly for those who are not familiar with the science of taxonomy.

However, while photographing the plants it became clear that the subjective perception of colour would play an important role in determining which plants were placed in each colour chapter. As a result we decided to amalgamate the purple and blue plants into one chapter when it became abundantly clear that one person's blue was another person's purple! Furthermore,

where plants have tiny insignificant flowers that are, for example, technically green yet for the most part the plant appears red, we have opted to place it in the red chapter on the basis that the main principle of this book is identification based on the first impression of each plant.

Scale

In order to give some impression of the scale of each plant the size has been described approximately in the text, for example as twice life size, life size or half life-size.

The text

It has never been our intention to produce a traditional botanical field guide. Our vision has been to create a book that makes wild flowers accessible to those who have little or no botanical knowledge and to provide those with more experience with information beyond that needed to identify wild flowers in their natural habitats. The concept of this book has always been about instilling enthusiasm into anyone with even the most vague interest in wild flowers, while hopefully increasing their knowledge.

In the text a brief description of each plant is given, which includes botanical (Latin) and common names, the type of plant (perennial, biennial, annual, climber or sub-shrub), height, scale, habitat, distribution through northern Europe (which includes Britain and Ireland except where stated) and flowering period. This information is supplemented by a description of each plant's structure (flowers, leaves, stems, roots and where significant the fruits and seeds), facts on pollination and seed distribution, folklore and anecdotes, and the origins of Latin names, British common names and some

LEFT The demure chickweed wintergreen (*Trientalis europaea*) possesses hidden talents for curing blood poisoning and mending wounds.

Wild-flower expert and photographic consultant to this book, Linda Laxton's nursery, British Wildflower Plants in Norfolk, stocks over 400 species of native wild flowers.

common names used in the other countries of northern Europe. Where possible, there are also suggestions on where and how to grow these plants in the garden, with recommendations of alternative garden cultivars and methods of eradication if the plant is considered a garden weed.

Needless to say, none of this information could have been given without recourse to the wealth of reference books devoted to wild flowers, most of which have been labours of love and taken many years to research and write, and all of which are listed in Further Reading (page 249). This book aims to bring a little of this information together in one place.

Height and spread

We have opted to provide dimensions for the height only of each plant. Spread has been omitted because it is more variable than height, being influenced mainly by the soil conditions in which the plant is growing. For example, a 50cm high plant may be growing in good soil allowing it to become much more branched and bushy, while the same plant growing in poor soil can be the same height but single-stemmed and with no branches.

IMPORTANT NOTE

Wild Flowers of Britain and Ireland does not intend to encourage the collection of wild flowers from their natural habitats under any circumstances. Modern photographic equipment provides by far the best and most rewarding means for recording your finds in situ without compromising populations of wild plants. If you own an SLR camera, invest in a macro lens or an extension tube at relatively little cost for close-up shots. Pocket digital cameras are lightweight and, without film costs to consider, they are economical and provide instant results. Otherwise these days most of us own a mobile phone and many models are now equipped with cameras – some even have a zoom option that allows close-up shots of the smaller wild-flower species. A x10 hand lens will allow you to scrutinise the detail of each plant up close.

A brief history of wild flowers in Britain and Ireland

Around 15,000 years ago Britain, Ireland and northern Europe was one landmass and in the grip of the Ice Age, which nearly emptied the landscape of vegetation. When the ice finally retreated, herbaceous plants began a rapid comeback, arriving in Britain from Europe via the tract of land that connected them. As Britain and Ireland were separated from Europe by the rise in sea level, the movement of wild-flower seed from one region to another was limited. Fewer new species of wild flowers could establish themselves and instead slower-growing shrubs and trees formed dense primary or ancient woodland, which shaded out most of the herbaceous plants.

Neolithic man was the first to fell these immense tracts of primary woodland to make way for crops and livestock. Consequently areas of land were opened up again and exposed to the light, which allowed plants, such as the bluebell (page 170) and primrose (page 84), to successfully re-establish themselves. Subsequent generations of human settlers continued to fell trees and cultivate the land. For example, in Britain the Celts, who grew corn and raised cattle, were responsible for creating the semi-natural habitats where some of our most familiar wild flowers such as fat hen (page 217), grow today.

Many of the plants of Britain and Ireland now perceived as native wild flowers were introduced from the rest of Europe or the New World intentionally for medicinal, culinary and garden use. The Romans brought us potherbs such as Alexanders (page 83) and cornfield plants including the now rare corncockle (page 152) and corn marigold (page 103). Pineappleweed (page 111) was introduced from North America in the 19th century and, distributed on shoe soles and car tyres, has colonised tracks and paths where it tolerates trampling. Other wild flowers arrived unexpectedly in seed form among goods shipped from abroad. Today several garden escapees, such as the flamboyant monkey flower (page 123) from North America, have successfully survived beyond the garden wall and, conversely, many wild flowers have been cultivated as garden plants – crane's-bill geraniums and roses are particularly obvious examples.

ABOVE Harbinger of spring, the primrose (*Primula vulgaris*), or 'first rose', flourishes in the shelter of hedge-banks and under deciduous trees. Its simple beauty is not its only attribute – it has myriad medicinal uses as well as edible flowers.
LEFT Now a familiar plant of coastal wasteland, Alexanders (*Smyrnium olusatrum*) is an ancient potherb, cultivated by the Romans for its celery-like stems.

Wild flowers and their uses

For centuries man has exploited wild plants as sources of food, for medicine and for socio-economic purposes. Some wild flowers have become important food crops for both man and livestock. Red clover (page 160), for example, is grown both as a fodder crop and to enrich the soil through its nitrogen-fixing abilities. Other wild flowers, such as sea kale (page 51), reflexed stonecrop (page 112) and rock samphire (page 104) were harvested, occasionally at great human cost, as delicacies. Then there are those wild plants that are the ancestors of modern varieties of fruit and vegetables such as wild cabbage (page 83), which is said to be the progenitor of modern varieties of cabbage, Brussels sprouts and cauliflower. And dye plants, such as madder (for red), woad (for blue) and weld (for yellow), were all grown on a commercial scale until they were superseded by synthetic dyes.

Many wild flowers were bestowed with unlikely medicinal properties on the basis of the 16th-century philosophy known as the Doctrine of Signatures, thanks to their superficial resemblance to parts of the human body. Eyebright (page 60), for example, was thought to resemble the eye and as a result was meant to possess properties that cured eye ailments. Other wild flowers, however, had genuine medicinal properties, which were documented by famous herbalists such as John Gerard and Nicholas Culpeper. Some have since been exploited, developed and even synthesised in modern medicines – primrose (page 84) is still used in expectorants against bronchitis and common evening primrose (page 95) remains an accepted cure for pre-menstrual tension and symptoms of the menopause.

LEFT Sea kale (*Crambe maritima*) had been harvested from its coastal environs for centuries before it was cultivated as a vegetable.

Steeped in magic and mystery, mistletoe (*Viscum album*) has been intrinsic to religion and ritual for hundreds of years.

Throughout history, wild flowers have had fundamental roles in both religion and magic. In France goat's beard (page 95) is closely associated with the Catholic festival, St John's Eve. Countless wild flowers were woven into legends about fairies and elves; witches and the devil were regularly kept at bay by flowers hung over doorframes and, in Ireland, bunches of marsh marigold are said to protect fertile cattle from witches and fairies. Many wild plants are historically bound up with love and romance – the drowning knight in the romantic tale of the forget-me-not is particularly poignant (page 168), plucking a daisy's petals one by one will tell whether one's lover is true and, of course, Christmas wouldn't be complete without a kiss under the mistletoe (page 209).

RIGHT Men risked their lives to harvest rock samphire (*Crithmum maritimum*) from cliff-tops. The leaves were pickled and the stems cooked and smothered in butter.

Wild-flower habitats

How and where wild flowers grow is influenced by variations in soil, sunlight and climatic conditions; these in turn determine a wild flower's habitat.

Grassland
Grassland appears at both high and low altitudes and on a variety of soil types from

Wild strawberry (*Fragaria vesca*, page 36)

chalk to clay. The type of grassland and the number of wild-flower species it contains is also determined by the presence or lack of fertilisers. Those areas that have been fertilised to increase the level of nutrients tend to have a poorer diversity of wild flowers, while impoverished soils tend to have a richer diversity of wild flowers.

Chalk grassland
Chalk grassland is an ancient semi-natural habitat, which is the result of woodland clearance to create grazing for livestock. The soil is generally shallow and alkaline, and supports a rich variety of wild flowers. Many species are low growing, such as plantains (pages 43 & 217) and wild strawberry (pages 36), adapted to tolerate grazing by livestock (and rabbits) with flat

basal rosettes of leaves. Exposed to full sun, chalk grassland plants flower from late spring through to late summer.

Meadows
Meadowland is a semi-natural habitat, also resulting from ancient woodland clearance. Often exposed to full sun, spring meadows and summer hay meadows support a vast diversity of perennial wild flowers, such as bugle, lesser stitchwort, common knapweed and lady's bedstraw and associated animals and insects (see chapter on Gardening with Wild Flowers, page 240).

Wet meadows exist near lakes and rivers and are home to plants such as ragged robin (page 127) and snake's-head fritillary. Depending on how the land is managed, ground-hugging plants such as daisies (page 36) may be present if the grass is mown regularly. Where the meadow is wetter, it is more likely to become marsh.

Bugle (*Ajuga reptans*, page 174)

Bluebell (*Hyacinthoides non-scripta*, page 170)

Woodland
Few tracts of the primary or ancient woodland described above remain. Most of the woodland we recognise today has been managed by man for thousands of years, whether as coppiced woodland or through selective felling. There are two types of woodland: coniferous and deciduous. Coniferous woods, which tend to grow on acid soil and consist mostly of pine trees in Britain and Ireland, cast deep shade, and as a result, fewer wild flowers grow there. Deciduous woodland consists mainly of oak, ash, beech and birch trees. Soil types range from shallow and dry acid and alkaline soils to deep, clay soils. A wide range of plants benefit from the light that filters down to the woodland floor, with spring-, autumn- and winter-flowering plants coming into flower when the shade is at its lightest. Plants of deciduous woodland include wood anemones (page 31), bluebells (page 170) and snowdrops (page 31).

RIGHT Wood anemone (*Anemone nemorosa*, page 31)

Bog orchid (*Hammarbya paludosa*, page 226)

Common glasswort (*Salicornia europaea*, page 225)

Common sea lavender (*Limonium vulgare*, page 163)

Heath and acid bogs

Most heath is semi-natural, having been established as a result of man's agricultural activities. The soil tends to be very acid and the variety of plants that can tolerate such a low pH is limited. Plants that do thrive in these conditions, however, include the heathers (pages 198 & 206). Areas of heath where moisture cannot drain away turn into acid bog, which is often home to some of our most unusual wild flowers such as the sundews and elusive species of orchid.

Fens

Fens are waterlogged areas of land, rich in nutrients and usually on alkaline soil. If they are regularly maintained, either by mowing or cutting, to prevent them from being invaded by shrubs and trees, they are often home to a rich variety of summer-flowering herbaceous perennials.

Hedgerows

Hedgerows are man-made habitats and in Britain many were planted as boundaries to enclose Anglo-Saxon fields or are remnants

LEFT Cow parsley (*Anthriscus sylvestris*, page 32)

of woodland deliberately left as wind breaks. In Britain the Parliamentary Land Enclosures Acts of the 18th and 19th centuries enforced the planting of hedges to enclose pasture, fields and wasteland. Plants such as cow parsley (page 32) and garlic mustard (page 31) thrive in the shelter provided by hedges, while they support climbing plants such as old man's beard (page 79) and honeysuckle (page 96). The orientation of the hedge determines the type of plants that will colonise in the microclimate at its base, for example, sun lovers thrive on the south side of a hedge while shade-tolerant plants flourish on the north side (see chapter on Gardening with Wild Flowers, page 245).

Salt marshes

Salt marshes are coastal habitats that succumb to the twice-daily tidal comings and goings of the sea. The wild flowers that succeed in this habitat have adapted to tolerate high levels of salt in both the water and the air, for example, glasswort (page 225) and seablites (page 163). Some plants avoid the tides and colonise the upper reaches of the salt marsh where succulent leaves, papery petals and fine hairs are all devices that allow them to tolerate the

drying and salty winds. Sea lavender (page 163) is a spectacular example.

Beaches and sand dunes

Both sandy and pebbly beaches and sand dunes have little soil, are poor in nutrients and are exposed to strong coastal winds. However, these habitats still support a varied if limited range of wild flowers, which have adapted to survive their hostile conditions by possessing deep penetrating roots or succulent blue-grey foliage, for example. Plants that thrive in coastal conditions include sea holly (page 201), viper's bugloss (page 176) and edible sea kale (page 51).

Yellow-horned poppy (*Glaucium flavum*, page 103)

Freshwater ponds, rivers and streams

The range of wild flowers found in and around freshwater habitats is determined by several factors: the depth of water, the speed of the current, the composition of the bottom – i.e. whether it consists of sand, stones or mud – the quality of the water and the water's acidity (Rose, 1981) – water lilies, for example, will only grow in water that is unpolluted. Plants growing near or in water may be divided into oxygenators, plants that grow under the surface of the water and supply oxygen; floating plants, which have floating leaves and create shade that keeps the water cool; emergents, which grow with their roots under the water and their leaves above the surface, and marginals, which grow in the damp soil at the water's edge (see chapter on Gardening with Wild Flowers, page 243).

Common water crowfoot (*Ranunculus aquatilis*, page 40)

Cultivated land, wasteland, roadsides and railways

Land disturbed by man's activities seems an unlikely and even hostile environment for wild flowers to flourish, but in these habitats some of the most beautiful and diverse types of wild flower grows. For example, cornfields, where the soil is

LEFT Natural wetlands are a threatened habitat.

Hemp agrimony (*Eupatorium cannabinum*, page 143)

ploughed annually, are typically populated by cornfield annuals such as cornflower, corncockle (page 152), field poppy (page 160) and corn chamomile (*Anthemis arvensis*). On wasteland where the soil and vegetation are left unmanaged, persistent perennials such as stinging nettle and bindweed are familiar plants. On bare soil, annuals quickly colonise the area, often producing hundreds of thousands of seed in one season to guarantee their survival.

Transport infrastructures such as roads, railways and canals are often responsible for the successful spread of some wild flowers introduced from abroad. In Britain, the classic example of this type of distribution is Oxford ragwort (*Senecio squalidus*), which travelled from Italy to Oxford, where it was first recorded in the 18th century, and later established itself countrywide via railway embankments. Similarly, Himalayan balsam (see page 148) has spread successfully along the network of canals.

RIGHT Rosebay willowherb
(*Chamerion angustifolium*, page 139)

About the plant

ALTERNATE LEAVES

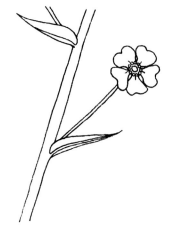

BRACT

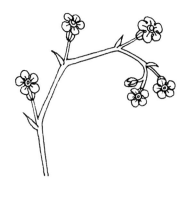

CYME

Achene A dry fruit that contains one seed and does not split.

Alien A plant that is not native to Britain or Ireland but has been introduced, either deliberately or accidentally, by man or through other means, from elsewhere.

Alternate Leaves arranged alternately on either side of the stem.

Annual A plant that germinates, flowers, fruits, sets seed and dies in one season.

Anther The part of the stamen where the cell divisions that form pollen grains occur.

Axil The angle between leaf and stem.

Basal Leaves that arise at the base of the stem.

Berry A fleshy fruit that contains several seeds.

Biennial A plant with a two-year life cycle: the seeds germinate in the first year, and flowers and fruits appear in the second.

Bract A leaf-like organ that appears immediately below a flower.

Bulb An underground storage organ consisting of a short stem and a bud enclosed in fleshy leaves.

Bulbil A small bulb-like organ that appears above ground from the leaf axil or bract.

Calyx All the sepals of a flower.

Chlorophyll Green pigment in plants, responsible for trapping sunlight to make food material.

Cleistogamous Flowers that do not open and are self-pollinating.

Compound leaf A leaf divided into several leaflets.

Corm A short vertical underground stem that acts as a storage organ for food.

Corolla All the petals of a flower.

Cultivar A variety of plant produced by selective breeding for horticultural use.

Cyme An inflorescence where a flower terminates the growth of each branch which may bear flowers on one or both sides.

Dioecious Where male and female flowers appear on separate plants.

Disc florets The central florets of a flower head that forms a disc shape, usually only in the daisy family (Asteraceae or Compositae).

Escapee Cultivated plants that have escaped into the wild.

Family A grouping of plants classified according to the similarity of their flowers and fruit and which can contain one or many genera, e.g. Ranunculaceae contains buttercups (*Ranunculus*) and monkshood (*Aconitum*).

Filament The stalk of the stamen.

Floret An individual flower of a flower head.

Flower head Where several individual flowers are held together in one group.

Form A minor variation on the normal form of a species. Variations can occur in the leaves, flowers, structure or habit of a plant.

Fruit The structure that contains the seeds, which is the fertilised ovary.

Genus (Genera) A grouping of one to many closely related species, each with similar features and denoted by the first part of the Latin name, e.g. ***Ranunculus repens*** and ***R. bulbosus***. Genera do not usually interbreed.

Gland A usually rounded structure that appears on the surface of a plant that contains and secretes a liquid.

Habitat The natural environment where a plant grows in the wild.

Hybrid The offspring produced naturally or artificially from two genetically different plants.

Inflorescence All the flowering branches of a plant.

Introduction A non-native plant to Britain and Ireland that has been brought in from another country.

Involucre Flower bracts that form a collar around the base of a group of flowers.

Keel A sharply folded edge, like the keel of a boat, often used to describe the lower petal in flowers belonging to the pea family. Some leaves are also keeled.

Lanceolate Leaves that are almost elliptical or lance-shaped.

Latex Milky juice produced in stems and leaves of some plant families, e.g. Euphorbiaceae and Asteraceae.

Leaflet One part of a compound leaf.

Linear Leaves that are long and narrow, with parallel sides.

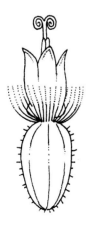

DISC FLORETS

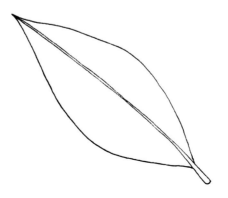

LANCEOLATE LEAF

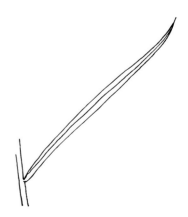

LINEAR LEAF

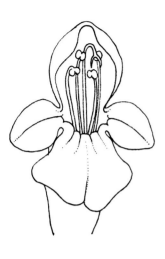

LIP OF COROLLA

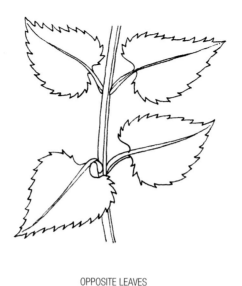

OPPOSITE LEAVES

PALMATE LEAF

Lip A petal that is obviously divided from the rest of the petals, with its parts often fused together.

Lobed Leaves or petals that are divided but not entirely separated.

Midrib The central main vein of a leaf.

Monoecious Where male and female flowers appear on the same plant.

Mycorrhiza Fungae that live in association with the roots of a plant and supply nutrients and moisture.

Native A plant that occurs naturally in Britain and Ireland.

Naturalised A non-native or introduced plant that appears to grow wild.

Nectar Sugary liquid produced in the nectaries of flowers and collected by insects.

Nectary A modified petal or stamen that contains and secretes nectar.

Node Where leaves arise on a stem.

Nodule A swelling on leaves or roots. In members of the pea family they contain bacteria that 'fix' nitrogen from the air making it available to the plant as a nutrient.

Oblong Petals and leaves that are rectangular but with rounded ends.

Opposite leaves Leaves arranged in pairs on opposite sides of the stem.

Oval Petals and leaves with rounded ends.

Ovary The part of the female organ that contains the ovules.

Ovule The organ that contains the egg, which develops into a seed once fertilised.

Palmate Leaves that look hand-like, consisting of more than three lobes.

Panicle An inflorescence with many branches and sub-branches.

Parasite A plant that relies entirely on another plant for nutrients and moisture.

Pedicel The stalk of an individual flower.

Perianth All the petals and sepals – a term used when they cannot be differentiated from each other.

Perennial A plant that can live for more than two years and flowers each year.

Petals The parts of a flower inside the sepals; some flowers have no petals.

Petiole A leaf stalk.

Photosynthesis The chemical process by which light energy (absorbed by chlorophyll) is used to produce plant food from carbon dioxide from the air and water.

Pinnate A compound leaf where leaflets are arranged in opposite pairs and which may or may not have a leaflet at the top.

Pistil The female part of a flower.

Pollen Tiny particles produced by the anthers and containing the male gametes or pollen grains.

Pollinia A mass of pollen grains.

Raceme A spike-like inflorescence where each flower is held on its own stem.

Ray A spoke-like stalk that radiates from the stem in an umbel.

PANICLE

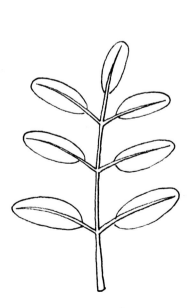

PINNATE LEAVES

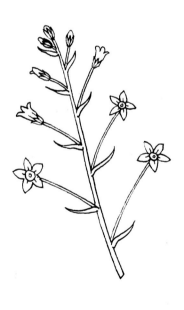

RACEME

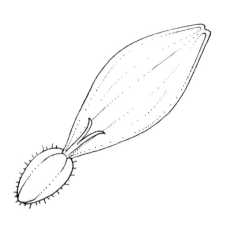

RAY FLORET

SPIKE

TRIFOLIATE LEAF

Ray floret One of the small flowers that form a ring around the disc floret.

Reflexed Petals or leaves that are bent backwards or downwards.

Rhizome An underground horizontal stem that is usually swollen and fleshy.

Rootstock The underground part of a plant, which anchors the plant to the soil and absorbs nutrients and moisture.

Rosette Leaves at the base of a plant with a flower stem arising from the centre.

Runner A horizontal stem that grows above ground and roots at the nodes and at its tip.

Sap Liquid contained in the leaves and stem.

Saprophyte A plant that contains no chlorophyll and so cannot make food materials, so lives on decaying organic matter.

Scape A flower-bearing stem with no leaves.

Semi-parasite A plant that relies partly on other plants for nutrients and moisture.

Sepal The usually green or brown outer part of the flower that protects the flower bud. Collectively known as the calyx.

Spadix A dense spike-like flower head enclosed in a large bract or spathe.

Spathe A single large, sheath-like bract that encloses the spadix in some plants.

Species The sub-divisions of a genus, which are denoted by the second part of the Latin name, e.g. *Ranunculus **repens*** and *R. **bulbosus***. Species are similar in appearance and can interbreed.

Spike An inflorescence whose flowers are attached to the stem without stalks.

Spur A hollow tubular structure that protrudes from the flower and usually contains nectar.

Stamen The male reproductive organ of a flower, which contains the anther and filament.

Stigma The upper part of the female reproductive organ of a flower, which receives the pollen.

Stolon A creeping stem that grows above or below ground.

Style The part of the female reproductive organ of a flower that joins the stigma to the ovary.

Subspecies (abbrev. subsp.) A plant that has distinct variations from the normal form of a species by virtue of its geographical distribution and is genetically different enough to breed true, but not to be a species in its own right.

Synonym (abbrev. syn) Another name for a plant that is not generally accepted or used.

Taproot A deep-growing main root, which is often swollen and stores food.

Tendril A curled modified leaf, leaflet or stem used for support when climbing.

Tepal A perianth segment that looks similar to all the others.

Trifoliate leaf A leaf consisting of three distinct parts.

Tube The narrow, cylindrical part of the corolla or calyx.

Tuber A swollen underground stem or root where food is stored.

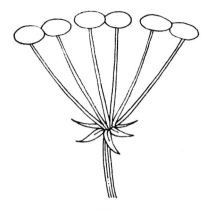

UMBEL

WHORL

Umbel An umbrella-like inflorescence where the stalks radiate from one point at the tip of the stem.

Variety (abbrev. var.) A plant that differs from the normal form of a species in its botanical structure.

Whorl A cluster or ring of three or more leaves or flowers, which appear at the same level on a stem.

PARTS OF A PLANT

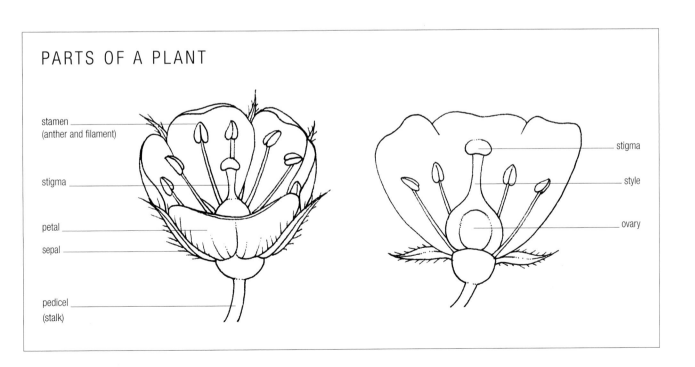

stamen (anther and filament)
stigma
petal
sepal
pedicel (stalk)
stigma
style
ovary

The Plants

Wood anemone, windflower [1]

Anemone nemorosa (Ranunculaceae)

Perennial. Height Up to 30cm. **Habitat** Woodlands and hedgerows. **Distribution** All northern Europe except Faeroe Is and Iceland. **Flowers** March–May. **In gardens** Moist, well-drained soil in partial shade; woodland gardens and damp shady borders. **Image size** Life-size.

It is a sure sign that spring has sprung when wood anemones burst into flower, carpeting deciduous woodland with drifts of exquisite nodding white flowers. The stems grow from creeping rootstocks, and are thin and flexible. What look like petals are actually sepals. They often have a pink tinge on the back, and pink- and blue-flowered varieties are also found; each flower may have up to 70 stamens in the centre. The flowering stems have three deeply cut three-lobed leaves in a whorl, part-way up. Similar leaves, growing directly from the root on long stalks, appear after flowering.

The wood anemone's beauty has made it popular for gardens, and there are many delightful cultivars. 'Flore Pleno' has white double flowers, 'Blue Bonnet' is deep blue and 'Bracteata Pleniflora' has semi-double white flowers surrounded by a leafy ruff.

Spring beauty [2]

Claytonia perfoliata (syn. *Montia perfoliata*; Portulacaceae)

Annual or perennial. Height Up to 30cm. **Habitat** Wasteland. **Distribution** Introduced to Belgium, Britain, Ireland, France, Germany and Holland. **Flowers** May–July. **In gardens** Shady wild gardens, and as a salad crop; a weed on sandy soil. **Image size** Life-size.

Spring beauty is an unusual-looking plant. The dainty white flowers seem to grow from a saucer-like pair of stalkless stem leaves that are joined together at their base. They are held in a loose cluster and have two sepals and five white petals, each with a notch at the tip. The basal leaves are long-stalked, with an oval pointed blade. All the leaves are slightly succulent, or fleshy.

Barren strawberry [3]

Potentilla sterilis (Rosaceae)

Perennial. Height Up to 15cm. **Habitat** Open woodlands, scrubland and hedges on chalky soils. **Distribution** All northern Europe to southern Sweden. **Flowers** February–May. **Image size** Life-size.

If you love wild strawberries (p36), it can be a disappointment to come across a patch of the barren strawberry. At first sight it looks like a perfect imitation, with its creeping rootstock, long-stalked three-lobed leaves and white flowers, but it belongs to a different genus altogether. On closer examination, the rootstock is not the same vigorous runner produced by the wild strawberry, the leaves have a bluish-green tint, and the flowers have gaps between the petals. Instead of succulent red berries, the fruits are yellowish and never fleshy.

Snowdrop [4]

Galanthus nivalis (Amaryllidaceae)

Bulbous perennial. Height Up to 25cm. **Habitat** Woods, hedgerows, churchyards, meadows and stream-banks. **Distribution** Probably introduced. Naturalised throughout northern Europe as far north as southern Sweden. **Flowers** January–March. **In gardens** Moist, rich soil in partial shade. Under trees or in grass. **Image size** One-third life-size.

Among the best-known and best-loved wild and cultivated flowers, snowdrops are eagerly awaited every year. As the grey-green leaves and white flowers force their way through the soil, they are certain indicators that spring is on the way. Snowdrop flowers have three pure white spreading sepals, which are about twice as long as the three inner petals – which are tipped with green. Snowdrops rarely set viable seed and seem to spread mainly vegetatively; they are best planted in spring when in leaf. Two other species, *Galanthus elwesii*, with broader leaves, and *G. plicatus*, with folded leaf edges, are also easy to grow in gardens.

Garlic mustard, Jack-by-the-hedge [5]

Alliaria petiolata (Brassicaceae or Cruciferae)

Biennial. Height 20–120cm. **Habitat** Hedgerows, wood margins, roadsides and wasteland. **Distribution** All northern Europe except Faeroe Is and Iceland. **Flowers** April–June. **Image size** Half life-size.

Crush the leaves or sniff the roots and you will quickly discover why this plant is called garlic mustard: it is the only member of the cabbage family to smell of garlic. The 3–5mm flowers have four pure white petals. Long thin fruits up to 7cm long follow and in early summer, these unzip as they dry, releasing black seeds. Glossy, heart-shaped leaves with toothed edges huddle at the base of the plant in a rosette; farther up the stem they are more angular. The culinary uses of garlic mustard were first recorded in the 17th century, when it was used to flavour salt fish; today it has graduated to gourmet spring salads.

Ramsons, wild garlic [6]

Allium ursinum (Liliaceae)

Perennial. Height 45cm. **Habitat** Damp woodland and hedge-banks on chalky soil. **Distribution** All northern Europe to Denmark. **Flowers** April–May. **Image size** Life-size.

You will probably smell ramsons before you see it – the bright green elliptical leaves smell strongly of garlic, especially if crushed. Several British place names – such as Ramsholt in Suffolk and Ramsbottom in Lancashire – come from the Old English name, *hramsa*. Ramsons spreads prolifically and its swathes of white starry flowers, 12–20mm across in loose umbels, regularly co-star with bluebells (p.170) in one of nature's best late-spring spectacles. The bulbs are too small for culinary use, but the leaves can be used in cooking or in salads.

Greater stitchwort [1]

Stellaria holostea (Caryophyllaceae)

Perennial. Height Up to 60cm. **Habitat** Woodland and shady hedgerows. **Distribution** All northern Europe except Iceland and Faeroe Is. **Flowers** April–June. **Image size** Two-thirds life-size.

A magnet for butterflies, bees, moths and beetles, the greater stitchwort is a must for wild-flower gardens. The square stems grow tall, lanky and rather brittle, and need the support of neighbouring plants. The rough edges of the long, tapering leaves help to give it a grip. (The related wood stitchwort, *S. nemorum*, which colonises damp woodland, has oval leaves.) The 18–30mm flowers are in loose clusters and have five pure white deeply cleft petals, each twice as long as the stamens; they have been called 'poor-man's buttonhole'. The common name refers to its use in relieving a 'stitch'. Its seed capsules burst with an audible 'pop'.

Wood sorrel [2]

Oxalis acetosella (Oxalidaceae)

Perennial. Height Up to 15cm. **Habitat** Woodland and hedgerows. **Distribution** All northern Europe. **Flowers** April–June; August. **In gardens** Humus-rich soil in sun or partial shade; woodland gardens and shady borders. **Image size** Two-thirds life-size.

The familiar clover-shaped foliage, which folds up in the late afternoon, makes wood sorrel one of several contenders for the shamrock of St Patrick. The solitary flowers emerge direct from the creeping roots and have five white petals with purple or violet veins; they droop at night or in rain for protection. The spring flowers attract plenty of pollinators but produce very little seed. The main seed production comes from short-stalked flowers in high summer; these never open and are self-pollinating.

The leaves have the sharp flavour of oxalic acid – hence the botanical name (from the Greek for 'sour') – and were added in small quantities to spring salads. Too much oxalic acid is poisonous.

Three-cornered garlic [3]

Allium triquetrum (Liliaceae)

Perennial. Height Up to 45cm. **Habitat** Hedgerows, woodland and wasteland. **Distribution** Introduced to Britain and Ireland. **Flowers** April–June. **Image size** Two-thirds life-size.

Similar to a white bluebell, the three-cornered garlic looks at first good enough to plant in the garden. But this is not recommended, because it is extremely invasive. The elongated stems and leaves are all triangular in cross-section. The drooping flower heads are one-sided, consisting of 3–15 bell-shaped flowers, with a green line down the middle of each segment. Since being introduced to Britain in the mid-1700s, the three-cornered garlic has spread widely.

White deadnettle [4]

Lamium album (Lamiaceae or Labiatae)

Perennial. Height Up to 60cm. **Habitat** Roadsides, hedges and wasteland. **Distribution** All northern Europe. **Flowers** April–December. **Image size** Two-thirds life-size.

It is almost impossible to tell the white deadnettle's leaves from those of the unrelated stinging nettle (p.218), but if the plant has large white flowers you can be sure they will not sting you. The flowers have a lobed bottom lip and hooded top lip, and have evolved to attract pollinating bees. The bottom lip acts as a landing pad for the bee, which helps itself to the nectar at the bottom of the flower tube. At the same time, it brushes against the stamens that dangle, protected by the hood, from the roof of the upper lip, and collects pollen which it carries to the next flower.

Cow parsley [5]

Anthriscus sylvestris (Apiaceae or Umbelliferae)

Biennial or short-lived perennial. Height Up to 1.5m. **Habitat** Hedge-banks, roadsides, wood margins and wasteland. **Distribution** All northern Europe except Faeroe Is and Iceland. **Flowers** April–June. **In gardens**
Well-drained soil in sun or partial shade; wild-flower meadows. **Image size** Two-thirds life-size.

The British name Queen Anne's lace aptly suits cow parsley's froth of small white scented flowers, which form showy ranks along verges and hedge-banks in late spring. It can be mistaken for similar-looking but poisonous relatives such as hemlock (p.59) and fool's parsley (p.68), and has been tarred with the same brush, sometimes being called Devil's meat. Yet it is harmless, and the finely divided leaves are edible. Bur chervil (*A. caucalis*) is an annual species with egg-shaped seeds covered in hooks, instead of the long, smooth seeds of cow parsley.

Bogbean [6]

Menyanthes trifoliata (Menyanthaceae)

Semi-aquatic perennial. Height Up to 30cm. **Habitat** Ponds, shallow lakes and wettest parts of fens and bogs. **Distribution** Scattered through most of northern Europe. **Flowers** May–July. **In gardens** Wet ordinary soil, best confined in container. Pool margins and bog gardens. **Image size** Half life-size.

Bogbean was named for its trifoliate bean-like leaves. Its rhizomes spread rapidly across open shallow areas of ponds or lakes, colonising large areas. The five-petalled whitish flowers are borne on spikes of up to 20 on each, opening from pink buds.

Meadow saxifrage [1]

Saxifraga granulata (Saxifragaceae)

Perennial. Height Up to 50cm. **Habitat** Meadows and other grassy places, and rocky habitats. **Distribution** Most of northern Europe except Faeroe Is and Iceland. Rare in Ireland. **Flowers** April–June. **In gardens** Well-drained but moist soil in full sun or light shade. Rock gardens or in grass. **Image size** Two-thirds life-size.

Come upon a clump of meadow saxifrage and you will soon realise why it was the first saxifrage to make it into gardens. The clusters of charming white flowers have five rounded petals suffused with green veins. They are held aloft on hairy, wiry stems above lobed, kidney-shaped leaves.

The plant was believed to have powerful curative properties by virtue of the tiny brown bulbils that grow in the angle between the stems and lower leaves. These were held to signify the plant's ability to cure gallstones but, in fact, are the method it uses to propagate itself. The species continues to be cherished as a popular garden plant.

Sea campion [2]

Silene uniflora (syn. S. maritima; Caryophyllaceae)

Perennial. Height Up to 30cm. **Habitat** Coastal shingle, rocks and cliffs; also mountains and lakesides inland. **Distribution** All northern Europe. **Flowers** June–August. **In gardens** Well-drained soil in sun. Coastal rock gardens. **Image size** Two-thirds life-size.

The low-growing sea campion straggles its way along coastal habitats forming mats of slender flowering and non-flowering stems. The 2cm flowers are similar to those of bladder campion (p.67) – pure white, with deeply notched petals and a distinctive inflated calyx – but there are far fewer in each flower cluster. The grey-green lanceolate leaves are arranged in opposite pairs and are slightly fleshy, helping the plant to tolerate drying winds. The garden cultivar 'Robin Whitebreast' (syn. 'Flore Pleno') has double flowers.

Sweet woodruff [3]

Galium odoratum (Rubiaceae)

Perennial. Height 45cm. **Habitat** Deciduous woodland on chalky soil. **Distribution** All northern Europe. **Flowers** May–June. **In gardens** Well-drained soil, partial shade. Woodland gardens, shady areas **Image size** Two-thirds life-size.

The creeping sweet woodruff is particularly abundant in beech and ash woods, where it thrives in dappled shade. The upright stems carry at their tip tiny funnel-shaped 4–7mm flowers on slender branches. They have only a faint scent, but this becomes more pronounced as the plant dries out, when it smells like new-mown hay. Like lavender, dried bunches used to be packed among linen to keep it smelling fresh and to discourage insects. Sweet woodruff is a pretty ground-cover plant for woodland gardens; in wildlife gardens it creates a haven for bees.

Danish scurvy grass [4]

Cochlearia danica (Brassicaceae or Cruciferae)

Annual or biennial. Height Up to 25cm. **Habitat** Sandy, rocky coasts and road verges. **Distribution** All northern Europe except Iceland and Faeroe Is. **Flowers** January–September. **Image size** Life-size.

Although a classic seashore annual, Danish scurvy grass has joined a select few – including lesser sea spurrey (p.151) – that have taken to the open road. Because of the large amounts of salt used on major roads in winter, some roadside verges have turned into salt turf. Danish scurvy grass has taken this opportunity to colonise areas it would never normally have reached. In other places, where salt is not used on the roads, it behaves itself and stays as a coastal plant.

It is a small plant that has four white or purplish-mauve petals. The heart-shaped basal leaves have long stalks, while the stem leaves have shorter stalks and are similar in shape to an ivy leaf. The lower leaves of English scurvy grass (*C. anglica*) are lobed.

Common whitlow grass [5]

Erophila verna (Brassicaceae or Cruciferae)

Annual. Height 10cm. **Habitat** Grassland, dunes, rocks and walls. **Distribution** All northern Europe except Faeroe Is. **Flowers** March–May. **Image size** Two-thirds life-size.

This diminutive plant thrives in habitats that are too inhospitable for most others, and is one of the first to bloom in early spring. The wavy stems grow from a flat rosette of spear-shaped toothed leaves, and support tiny white flowers. These have four notched petals and six deep yellow stamens. They are followed by seedpods, which contain two rows of seeds; when dry they turn opaque with an attractive pearly sheen.

Blinks [6]

Montia fontana (Portulacaceae)

Straggling annual or perennial. Height Stems up to 20cm long. **Habitat** Bare, usually wet lime-free habitats. **Distribution** All northern Europe. **Flowers** April–October. **Image size** Twice life-size.

Blinks produces tiny, insignificant white flowers on slim, branched stems which have a reddish tinge. It colonises places where the ground is regularly flooded, but these are now often drained to make way for crops. Yet it is a survivor and deserves respect as such. Its genus name commemorates Italian botanist Giuseppe Monti.

Horseradish [1]

Armoracia rusticana (Brassicaceae or Cruciferae)

Perennial. Height Up to 1.5m. **Habitat** Grassy habitats, and waste and cultivated land. **Distribution** Widely naturalised throughout northern Europe except far north. **Flowers** May–June. **In gardens** Moist, well-drained soil in full sun. Herb and vegetable gardens. **Image size** One-third life-size.

It is unclear when horseradish arrived in Britain and other parts of northern Europe from western Asia. However, it is thought to have been used medicinally before 1500 but not as a culinary herb until about 1750. Today its thick, white fleshy roots are grated and combined with soured cream to make a searingly hot sauce. It is a handsome plant, which escaped gardens to take root on waste ground and grassy habitats. Its large, toothed and lobed leaves grow on long stalks arising at the base of the stems; smaller leaves are arranged alternately up the flower stems and have shorter stems. The white four-petalled flowers are held in panicles but, in countries with a variable climate, there is no guarantee that plants will produce them every year.

Three-veined sandwort [2]

Moehringia trinervia (Caryophyllaceae)

Scrambling annual or short-lived perennial. Height Stems up to 40cm long. **Habitat** Woodland and other shady habitats on rich soil. **Distribution** All northern Europe except Iceland and Faeroe Is. **Flowers** May–July. **Image size** Life-size.

There are three groups of sandworts, all with undivided petals, unlike other white-flowered members of the family (such as pinks and carnations). One sandwort group has narrow leaves, the second short oval leaves, while three-veined sandwort has leaves that are oval, pointed, and with three prominent veins. The solitary flowers have five pointed sepals, which are longer than the five white petals. The round fruits shed seeds through six to eight valves.

Wild strawberry [3]

Fragaria vesca (Rosaceae)

Spreading perennial. Height 25cm. **Habitat** Dry grassland, woodland, hedgerows and roadsides. **Distribution** All northern Europe except Faeroe Is. **Flowers** April–July. **In gardens** Moist, well-drained soil, full sun or dappled shade. Herb gardens, containers, hanging baskets. **Image size** Two-thirds life-size.

Pure white 12–18mm flowers with bright yellow stamens precede the tiny red fruits, which dangle on long stems from the nodes of deeply toothed leaves. Wild strawberries were immortalised as symbols of eroticism in the 16th-century painting *The Garden of Earthly Delights* by Hieronymus Bosch, but to countless generations of children and adults they are simply among the sweetest fruits of summer. It is not now considered acceptable to transplant wild strawberry plants to gardens, but it is easy to raise them from seed.

We take the strawberry for granted as a staple ingredient of jams and summer puddings but tend to forget that it also has medicinal properties for the relief of diarrhoea and to soothe sunburn.

Scurvy grass [4]

Cochlearia officinalis (Brassicaceae or Cruciferae)

Biennial or perennial. Height 10–50cm. **Habitat** Salt marshes and grassy cliffs. **Distribution** All northern Europe except Iceland and Finland. **Flowers** April–August. **Image size** Two-thirds life-size.

The fleshy, kidney-shaped leaves of scurvy grass contain a sulphur-based oil, explaining their unpleasant odour and bitter taste. Nevertheless, the leaves were eaten raw by sailors of old to prevent scurvy on long voyages; it is now known that the leaves are also rich in vitamin C.

The white flowers, up to 10mm across, have the four petals typical of the cabbage family and bloom all summer long. Rounded seedpods follow.

Large bittercress [5]

Cardamine amara (Brassicaceae or Cruciferae)

Perennial. Height Up to 60cm. **Habitat** Damp meadows and woods, marshes and streamsides. **Distribution** All northern Europe except Faeroe Is and Iceland. **Flowers** April–June. **Image size** Two-thirds life-size.

The 12mm white flowers of large bittercress have the four rounded petals typical of the cabbage or cress family. However, the contrasting purplish stamens set it apart from its fellow species, whose stamens are yellow. The angular stems are little or not at all branched, and the flowers appear in loose clusters. The pinnate leaves consist of up to four pairs of leaflets with deeply toothed margins. They grow alternately up the stem, with (unlike many relatives) no basal rosette.

Daisy [6]

Bellis perennis (Asteraceae or Compositae)

Perennial. Height 2.5–15cm. **Habitat** Grassy habitats. **Distribution** All northern Europe. **Flowers** Year-round. **In gardens** Well-drained soil, full sun, partial shade. Wild-flower gardens, bedding. **Image size** Two-thirds life-size.

Everyone is familiar with the daisy, its bright yellow discs surrounded by a 15–30mm collar of white petal-like rays, tipped with red and flushed purplish underneath. The leaves grow in rosettes so flat against the ground that nothing else can grow under them. For generations, daisies have been strung into necklaces or plucked bald of their petals to predict the outcome of a love affair.

For those who prefer pristine lawns, they are a blot on the landscape to be dug out with a trowel or sprayed with spot weedkiller. For others, they are a welcome sight in early spring, and if you happen to find seven daisies that you can stand on at once, 'summer is come'.

Hoary cress [1]

Lepidium draba (syn. *Cardaria draba*; Brassicaceae or Cruciferae)

Perennial. Height Up to 90cm. **Habitat** Roadsides and wasteland. **Distribution** Introduced throughout northern Europe. **Flowers** May–June. **Image size** Life-size.

This untidy, weedy plant was introduced to Britain during the 19th century from the southern Mediterranean. The grey-green leaves clasp the upright, branched stems, which end in wayward clusters of small, white, cress-like flowers. The heart-shaped fruits are topped by a small beak, which hooks on to passing people and animals. Along with the plant's ability to produce root-buds, these enable it to advance rapidly throughout the countryside, particularly on roadsides.

Thale cress [2]

Arabidopsis thaliana (Brassicaceae or Cruciferae)

Annual or biennial. Height Up to 40cm. **Habitat** Walls, rocks and banks. **Distribution** All northern Europe except Iceland and Faeroe Is. **Flowers** April–May. **Image size** Life-size.

Thale cress is named after the 16th-century German botanist Johannes Thal. It has the distinction of being the first plant to have its genome (full set of genes) sequenced; it was chosen because it is easily cultivated and has a small number of chromosomes – 10, compared with humans' 46. The dainty flowers are small and white, with petals twice as long as the sepals. The basal rosette of foliage consists of elliptical leaves, while the stem leaves are narrow and more oblong. The long, slim fruits burst at the slightest touch when ripe, scattering their seed.

Lily-of-the-valley [3]

Convallaria majalis (Liliaceae)

Perennial. Height Up to 25cm. **Habitat** Dry woodlands, usually on chalk or limestone. **Distribution** All northern Europe except the extreme north, Iceland and Faeroe Is. **Flowers** May–June. **In gardens** Rich soil in sun or partial shade. Woodland gardens and damp shady borders. **Image size** Life-size.

Posies of deliciously scented lily-of-the-valley are redolent of early summer, and the plant has been cultivated for at least 500 years for use in cosmetics and perfumes. However, as a result, it can be difficult to tell true wild plants from well-established escapees; the former have smaller parts, while most garden forms are more vigorous.

Lily-of-the-valley is a creeping plant that eventually forms dense clumps of lush, oval foliage. The pure white flowers are carried on unbranched stems, with up to 12 flowers on each, all hanging to one side. The waxy petals and sepals are fused into a short bell-shaped tube. 'Albostriata' is a striking cultivar with vivid cream stripes along the length of the leaves.

Solomon's seal [4]

Polygonatum multiflorum (Liliaceae)

Arching perennial. Height 90cm. **Habitat** Chalky soil in woods and scrubland. **Distribution** Most of northern Europe; introduced to Ireland. **Flowers** May–June. **In gardens** Moist, well-drained soil, full or partial shade. Woodland gardens, shady borders. **Image size** Life-size.

The deeply ribbed, lance-shaped leaves of this most elegant of woodland plants appear alternately all the way up the arching stems. Clusters of tubular cream flowers, 9–20mm long and tipped with green, hang from each leaf node like tiny pearls; they are followed in the autumn by black fruits. The origin of the plant's name has long been debated, but is generally agreed to refer to the scars on the rootstocks. If you have a vivid imagination, they resemble the six-pointed seal (like the Star of David) of the Biblical King Solomon. Solomon's seal is one parent, along with *Polygonatum odoratum*, of the large-flowered garden hybrid *P.* x *hybridum*, a popular plant for shady borders and woodland gardens.

Mountain everlasting, cat's-foot [5]

Antennaria dioica (Asteraceae or Compositae)

Perennial. Height Up to 20cm. **Habitat** Mountains and moorland. **Distribution** All northern Europe except Faeroe Is and Iceland. **Flowers** June–July. **In gardens** Well-drained soil in full sun. Rockeries, alpine gardens or ground cover. **Image size** Life-size.

Everlastings are named because the flowers can be dried for long-lasting arrangements. Unusually in the daisy family, mountain everlasting has male and female flowers on separate plants. The male flower heads are small and daisy-like, but what look like ray florets are actually white bracts. The female flower heads are bigger and brush-like, surrounded by pink-flushed, woolly bracts – hence the name cat's-foot. The seeds have a parachute of long hairs.

Spoon-shaped leaves form a rosette at the base, while the stem leaves are narrower, with pointed tips. All the foliage is greyish-green, with white down underneath, so the whole plant looks greyish-white. The down reduces water loss from drying mountain winds. The plant spreads by leafy stolons (creeping stems), which root as they grow. It used to be quite common at lower altitudes, but intensive cultivation has destroyed many of its natural habitats. It is good in the garden or for cutting and drying.

Common water crowfoot [1]
Ranunculus aquatilis (Ranunculaceae)

Aquatic annual or perennial. Height Variable. **Habitat** Ponds, slow streams and ditches. **Distribution** Throughout northern Europe except Faeroe Is and Iceland. **Flowers** April–September. **In gardens** Informal water gardens and pools. **Image size** Life-size.

The water crowfoots – aquatic relatives of buttercups – are well adapted to life both in water and at the water's edge. The common species has two distinct types of leaves. The floating leaves are rounded and lobed, while below the surface fine feathery leaves are adapted for photosynthesis – which helps to oxygenate the water. Thread-leaved water crowfoot (*R. trichophyllus*) has only fine feathery foliage. Plump, solitary flower buds on fleshy stems open to reveal white saucer-shaped 15–20mm flowers, stained buttercup-yellow at the base of each petal. Happy in wildlife ponds, it is available from nurseries specialising in aquatic plants.

Marsh stitchwort [2]
Stellaria palustris (Caryophyllaceae)

Creeping perennial. Height Up to 60cm. **Habitat** Wet grassy places. **Distribution** All northern Europe. **Flowers** May–August. **Image size** Life-size.

Marsh stitchwort might spread on rounded creeping stems, but it has an upright growth habit. Blue-grey lanceolate leaves grow in pairs up the stems, while white flowers – each with five deeply lobed petals – are held aloft on slender stalks. They are pollinated by flies. A blend of stitchwort and acorns in wine was thought to relieve the pain caused by a 'stitch' in the side – hence the common name, which harks back to the 13th century.

Dame's violet, sweet rocket [3]
Hesperis matronalis (Brassicaceae or Cruciferae)

Biennial or short-lived perennial. Height Up to 1m. **Habitat** Grassland, wasteland and roadsides.

Distribution Introduced to all northern Europe. **Flowers** May–July. **In gardens** Moist, well-drained soil in full sun or partial shade. Wild-flower gardens and mixed borders. **Image size** One-third life-size.

A well-loved traditional cottage-garden plant, dame's violet has been cultivated for centuries and has escaped to hedgerows and roadsides. Yet it is an introduction from central and southern Europe. The flowers have four white or violet petals in loose panicles. They are fragrant at night – hence the name *Hesperis*, from the Greek for evening. But, according to the language of flowers, it also represents deceit, for there is no scent by day. The basal leaves are oval or oblong, those growing from the branching stem narrower, toothed and hairy.

Dame's violet is a must for wild gardens as a source of nectar for bees and butterflies. It is also a food plant for orange-tip butterfly caterpillars. For cut flowers, grow the double-flowered cultivars, such as *H. matronalis* var. *albiflora* 'Alba Plena' or 'Lilacina Flore Pleno'.

Sanicle [4]
Sanicula europaea (Apiaceae or Umbelliferae)

Perennial. Height 20–60cm. **Habitat** Deciduous woodland. **Distribution** All northern Europe except Faeroe Is and Iceland. **Flowers** May–August. **Image size** Half life-size.

Like miniature aerial pincushions, sanicle's compact umbels of pinkish- or greenish-white flowers are held aloft on wiry stems. Tiny insects pollinate the flowers, and the oval fruits, thick with hooked bristles, latch on to passing animals to guarantee seed distribution. The palmate leaves have reddish midribs and distinct toothed edges.

The name sanicle comes from the Latin *sanus*, meaning healthy, and the plant was reputed to relieve chest infections. In France and Germany the fresh juice, prepared from the leaves or roots and dispensed in tablespoon doses, was used to remedy profuse internal bleeding.

Pignut [5]
Conopodium majus (Apiaceae or Umbelliferae)

Perennial. Height 20–50cm. **Habitat** Grassland and arable land. **Distribution** Britain, Ireland, France and Norway. **Flowers** May–July. **Image size** Half life-size.

The 'nuts' of this relative of the carrot are its dark brown, knobbly underground tubers. They are edible and pigs were once trained to unearth them, like truffles. They often grow with creeping buttercups (p.115) and timothy, a grass cultivated for hay. However, the systematic loss of grass-land has made them rarer. In Britain it is now illegal to dig up the tubers without the landowner's permission.

Umbels of tiny white flowers bloom at the tips of ridged stems, which become hollow and brittle after flowering. The lower leaves have usually withered away by the time the flowers appear, and are very finely divided; the upper ones are less so.

White campion [6]
Silene latifolia subsp. *alba* (Caryophyllaceae)

Perennial or sometimes annual. Height 1m. **Habitat** Arable land and wasteland. **Distribution** All northern Europe. **Flowers** May–October. **Image size** Half life-size.

The white campion belongs to a genus of about 500 plant species named after the Greek god Silenus, merry god of the woods and foster-father to Dionysus (or Bacchus). By all accounts, Silenus spent most of his time drunk and covered in saliva. The white campion's deeply cut white flowers are held on stems covered in sticky hairs – hence the generic name. Similarly sticky, hairy leaves grow on stalks at the base of the stem, but towards the upper part of the stem the leaf-stalks are absent. Male and female flowers, 25–30mm across, appear on separate plants. Towards dusk, the female flowers give off a fragrance, which attracts pollinating moths.

hand from the base to ensure that the roots are completely removed. It is, of course, considered good luck to find a clover with four leaflets (or 'leaves') rather than three – best of all if it is found by accident.

Northern bedstraw [2]

Galium boreale (Rubiaceae)

Perennial. Height 20–40cm. **Habitat** Grassland, streamsides, rocky habitats and shingle. **Distribution** All northern Europe. **Flowers** June–August. **Image size** Half life-size.

A rather stiff plant, the northern bedstraw has dark green leaves, each with three main veins, arranged in whorls of four up the square stem. Clusters of white 3–4mm flowers appear at the branched tips. The name *Galium* is taken from the Greek *gala* (milk), and refers to the use of some related species to curdle milk. Other species were used to stuff straw mattresses, hence the common name. The flowers are pollinated by insects and the tiny brown fruits are covered with hooked bristles. By hitching a ride on the coat of passing animals, they better their chances of distributing the seed.

Brookweed [3]

Samolus valerandi (Primulaceae)

Perennial. Height 5–45cm. **Habitat** Chalky soil in damp and shady coastal habitats. **Distribution** All northern Europe except Faeroe Is, Iceland and Ireland. **Flowers** June–August. **Image size** Half life-size.

The genus name *Samolus* comes from two Celtic words: *san*, meaning health, and *mos*, meaning pig. It refers to the fact that the ancient Gauls regularly used this unassuming plant to cure illness among their herds of swine. The species name, *valerandi*, commemorates the 16th-century French botanist Dourez Valerand, who is credited with discovering the plant on a Greek island. Upright fleshy stems grow from a

White clover [1]

Trifolium repens (Fabaceae or Leguminosae)

Perennial. Height 50cm. **Habitat** Clay soils in grassy habitats. **Distribution** All northern Europe except the northernmost tip. **Flowers** June–September. **In gardens** Troublesome lawn weed. **Image size** Half life-size.

The triple bright green leaflets gave the rampant white clover its name *Trifolium*. Each is marked with a pale crescent and clings to the vigorous creeping, self-rooting stems by long, supple leaf petioles (stalks). The globular flower heads, 7–10mm across, vary from white to pale pink. Their sugary scent attracts pollinating bees; gently suck on the flower-ends and you will taste the sweet honey-like nectar.

With its ability to form substantial mats and clumps, the white clover is a persistent pest of garden lawns. Pull the plants out by basal rosette of spoon-shaped leaves and at their tip are spikes of delicate, cup-shaped flowers, 2–3mm across, arranged on alternate flower stems.

Hoary plantain 4
Plantago media (Plantaginaceae)

Perennial. Height 15cm. **Habitat** Grassland and wasteland. **Distribution** All northern Europe. **Flowers** May–August. **Image size** Two-thirds life-size.

The hoary plantain is very similar to the ribwort (see below), although it produces fewer leaves in flatter rosettes covered in curly hairs. The white flowers have pale lilac anthers and, unlike other species, give off a surprisingly strong fragrance that attracts pollinating bees. It is the only British plantain to be insect- rather than wind-pollinated. All plantains tolerate wear and tear by producing new growth at the very base of the plant. This explains why they tend to colonise paths, tracks and bridleways where few other plants can survive the pressures of consistent trampling.

Ribwort plantain 5
Plantago lanceolata (Plantaginaceae)

Perennial. Height 45cm. **Habitat** Dry meadows, wasteland and banks. **Distribution** All northern Europe. **Flowers** May–August. **In gardens** A pernicious weed. **Image size** Two-thirds life-size.

Like most plantains, this – one of the most common of all European plants – produces tiny flowers in dense upright spikes; generations of children have 'fired' them by looping and pulling on the wiry stems.

The flowers are brownish, and their most prominent feature is the disproportionately large creamy-white stamens, which appear in a ring around the middle of the spike; the overall effect has been perfectly described as 'a bad sombrero'.

It is a variable plant with deeply veined lanceolate leaves in tufted rosettes. Like all plantains, it is a great survivor, with deep roots and an ability to produce seed in large quantities. It survives trampling and mowing. In gardens it can be eradicated by using a trowel to dig up the deep roots or by applying an appropriate spot weedkiller.

Chickweed wintergreen [1]

Trientalis europaea (Primulaceae)

Creeping perennial. Height Up to 20cm. **Habitat** Damp grassland, heather moors and pine woods. **Distribution** All northern Europe except Ireland. **Flowers** June–July. **Image size** Two-thirds life-size.

Neither a chickweed nor a wintergreen, this plant throws up slim solitary stems from a creeping rootstock. They are upright and unbranched, and have a few alternate leaves below a whorl of shiny lanceolate leaves at the top. One or more anemone-like starry flowers emerge from the centre of the foliage. They are pure white and usually have five to nine petals.

When ripe, the rounded seed capsule splits into five segments. Chickweed used to be used to treat blood poisoning and wounds. The genus name *Trientalis* is from the Latin for one-third of a foot – the plant's usual height.

Pencilled crane's-bill [2]

Geranium versicolor (Geraniaceae)

Perennial. Height 45cm. **Habitat** Shady places. **Distribution** Introduced to Britain and Ireland. **Flowers** June–July. **In gardens** Well-drained soil, full sun or partial shade. Wild-flower gardens, woodland gardens, borders. **Image size** Two-thirds life-size.

A vigorous and successful garden escapee, the pencilled crane's-bill has exotic ancestry, being an introduction from southern Italy and the Balkans. Its white, funnel-shaped 25–30mm flowers, with deeply notched petals, have striking violet veins – hence the epithet 'pencilled'. They grow on hairy stems. The leaves consists of five deeply toothed lobes. In the wild, this species occasionally crosses with the pink-flowered French crane's-bill (*Geranium endressii*); the result is in the hybrid *G. x oxonianum*, or Druce's crane's-bill, which has white petals covered in a paler pink tracery of veins.

Common mouse-ear [3]

Cerastium fontanum (Caryophyllaceae)

Perennial. Height Up to 50cm. **Habitat** Grassy habitats, bare ground and shingle. **Distribution** All northern Europe. **Flowers** April–November. **Image size** Two thirds life-size.

The very variable common mouse-ear is one of about 60 species of mouse-ears, many of which are considered weeds where they colonise cultivated land. It is the soft, downy hairs on the stems and foliage that give them their common name. The leaves are lanceolate, without stalks, and arranged in opposite pairs along the trailing stems. Tiny white flowers are carried at the tip of each branched stem and have five deeply notched petals and sepals of equal length – a way of telling this species from the related and very similar field mouse-ear (*Cerastium arvense*), whose petals are twice as long as the sepals.

Mouse-ears produce curved, vaguely horn-like seed capsules, which release their seed only when the air is dry, ensuring that they will be carried on the wind.

Spring sandwort [4]

Minuartia verna (syn. *Alsine verna*; Caryophyllaceae)

Mat-forming perennial. Height Up to 15cm. **Habitat** Dry, rocky places and screes, often on limestone; mine spoils. **Distribution** Much of northern Europe except Holland, Denmark and Iceland. **Flowers** May–September. **Image size** Two-thirds life-size.

It may seem unlikely that such a pretty and dainty plant as spring sandwort could set up home on lead-mine spoil heaps, yet these scars on the landscape provide it with ideal growing conditions. The cushions of narrow, linear leaves have keeled undersides. At the stem tips grow loose clusters of five-petalled white flowers with contrasting purplish stamens and prominent sepals which are shorter than the petals. These distinguish this species from the otherwise similar fine-leaved

sandwort (*Minuartia hybrida*), which has sepals that are longer than the petals. The genus name *Minuartia* commemorates Juan Minuart, an 18th-century Spanish botanist; the species name *verna* means 'of spring'.

Mexican fleabane [5]

Erigeron karvinskianus (syn. *E. mucronatus*; Asteraceae or Compositae)

Perennial. Height Up to 50cm. **Habitat** Walls, pavements and stony ground. **Distribution** Naturalised in southern Britain, Ireland and parts of France. **Flowers** July–September. **In gardens** Fertile, well-drained soil in full sun or some shade. Walls, paving, containers and hanging baskets. **Image size** Two-thirds life-size.

Mexican fleabane forms a tangled web of thread-like stems, each with a 12–15mm daisy-like flower at its tip. The flowers have white, pink or pale purple ray florets with a mass of yellow disc florets at their centre. The small leaves are slender, oval and mostly three-lobed. The plant arrived from Mexico in the 19th century and remains (and is probably better known as) a popular garden plant. It does very well sprawling around on sunny walls – hence the alternative name wall daisy – or in the crevices of paving.

Scots lovage [1]

Ligusticum scoticum (Apiaceae or Umbelliferae)

Perennial. Height Up to 60cm. **Habitat** Rocky and shingly coastal habitats. **Distribution** Britain and Ireland eastwards to Germany and north to Scandinavia. **Flowers** July– August. **Image size** Two-thirds life-size.

Scots lovage is a rare member of the parsley, carrot and celery family, but is quite distinct from the herb lovage (*Levisticum officinale*). Like lovage it is celery-scented, but it bears closest resemblance to Alexanders (p.83). It throws up stout, grooved stems with a reddish tinge, whose branches are topped by umbels of tiny greenish-white flowers. The bright green leaves consist of between three and five oval leaflets with toothed edges. Like Alexanders, the base of the stem is edible and the leaves were eaten as a potherb, but it is too rare today for picking.

Sweet cicely [2]

Myrrhis odorata (Apiaceae or Umbelliferae)

Perennial. Height Up to 1.8m. **Habitat** Damp grassy habitats and waysides. **Distribution** France and Germany; widely naturalised elsewhere in northern Europe, except far north. **Flowers** May–July. **In gardens** Moist, well-drained soil in dappled shade. Herb and wild-flower gardens. **Image size** Two-thirds life-size.

Sweet cicely is an ancient potherb, relished for its aniseed flavour and scent. In some places it is known as myrrh, but it is no relation of the biblical myrrh. 'Cicely' is not a girl's name but simply a version of the Greek name *seselis*. It looks rather like cow parsley (p.32), with its frothy umbels of creamy-white flowers on upright grooved stems. But the distinctive scent gives it away, as does the lush mass of finely divided, fern-like leaves. The flowers are followed by ridged, somewhat gherkin-like fruits, which ripen to a dark brown. Sweet cicely has myriad culinary uses: boil and butter the roots, and eat them like parsnips; chop the raw leaves and add them to salads; or add the ripe seeds to fruit desserts.

Common gromwell [3]

Lithospermum officinale (Boraginaceae)

Perennial. Height Up to 1m. **Habitat** Hedgerows, scrub and woodland margins. **Distribution** All northern Europe except Faeroe Is and far north. **Flowers** May–August. **Image size** Two-thirds life-size.

Common gromwell is a stocky plant with stiff, upright, well-branched stems and lanceolate leaves with conspicuous veins. Small creamy or greenish-white funnel-shaped flowers are arranged in small clusters. Then come curious shiny white nut-like fruits, which were used in 16th- and 17th-century medicine to treat kidney and bladder stones and rheumatism. The name gromwell comes from French words meaning grey millet, and refers to the fruits. Its annual cousin, field gromwell (*Lithospermum arvense*, syn. *Buglossoides arvensis*) is similar but less branched.

Rock cinquefoil [4]

Potentilla rupestris (Rosaceae)

Perennial. Height 30–60cm. **Habitat** Rocky habitats in mountains. **Distribution** Northern Europe as far north as Norway and southern Sweden. **Flowers** May–June. **Image size** Half life-size.

Rock cinquefoil is extremely rare in Britain, where it grows only in a few locations in Wales and Scotland. It has a relatively short flowering period of only a month, so anyone who comes upon it in flower – anywhere in northern Europe – is lucky indeed. It is a hairy plant, with white 15–25mm cup-shaped flowers of rounded white petals with a coronet of bright yellow stamens at the centre. The pinnate basal leaves have five to seven toothed leaflets; the stem leaves are trifoliate and stemless. Its rarity could be due to its short flowering season and the fact that it self-pollinates – and not always successfully.

Ground elder [5]

Aegopodium podograria (Apiaceae or Umbelliferae)

Perennial. Height 30–90cm. **Habitat** Shady places. **Distribution** All northern Europe except the northernmost tip. **Flowers** May–August. **In gardens** A persistent weed. **Image size** Quarter life-size.

Ground elder's reputation as a thug is fully justified and long-held. Once the brittle, spreading rhizomes get established, it is almost impossible to eradicate. It was originally introduced by the Romans as a potherb and the fluffy spheres of tiny 2–3mm flowers – held on firm, grooved stems – can be attractive in shady garden spots. The foliage consists of deeply toothed leaves arranged in threes. Also known as goutweed, ground elder was thought to possess properties that soothe gout and achy joints. Banish it from the garden by forking it out of the ground, removing as much of the root as possible. If the problem persists, apply an appropriate weedkiller.

Hairy tare [6]

Vicia hirsuta (Fabaceae or Leguminosae)

Scrambling annual. Height Up to 80cm. **Habitat** Grassy habitats, woodland margins, roadsides, scrub and cultivated land. **Distribution** All northern Europe except Faeroe Is; naturalised in Iceland. **Flowers** May–August. **Image size** Half life-size.

As the name suggests, hairy tare produces seedpods that are coated in fine hairs, distinguishing it from smooth tare (*Vicia tetrasperma*) and slender tare (*V. tenuissima*). It has clambering stems and leaves that consist of up to ten pairs of small linear leaflets and end in branched tendrils. The white or purplish pea-like flowers are arranged in short racemes which grow on stalks from the leaf axils. Hairy tare, along with its relatives, has a reputation that harks back to biblical times for plaguing farmers. The wayward stems scramble through arable crops, clinging on by their tendrils and strangling anything in their way.

Ox-eye daisy, marguerite, moon daisy [1]

Leucanthemum vulgare (Asteraceae or Compositae)

Perennial. Height 20–70cm. **Habitat** Chalky or slightly acidic soils in grassy habitats, verges and scrubland. **Distribution** All northern Europe. **Flowers** May–September. **In gardens** Moist, well-drained soil, full sun, partial shade. Wild-flower gardens. **Image size** Two-thirds life-size.

Moon daisy or moonflower are apt names, as in the evening half-light the flowers really seem to glow in fields and churchyards. The large daisy heads, 50mm across, consist of a bright yellow disc surrounded by long white rays. They are supported on strong, wiry stems covered in fine hairs. The dark green leaves are coarsely toothed and grow at the base of the stem on short petioles.

In parts of Germany, France and Holland, the ox-eye daisy is closely associated with St John. In Germany the flowers were hung around doors to ward off lightning, and in some regions it is called *Gewitterblume*, or storm flower. It is recorded in several herbals as having a wide range of medicinal uses from soothing wounds to relieving bronchial catarrh. The garden marguerite or Shasta daisy, *L.* x *superbum*, is a related hybrid and is widely naturalised.

Hedge bedstraw [2]

Galium mollugo (Rubiaceae)

Sprawling and scrambling perennial. Height Up to 1m. **Habitat** Chalky soils in open woodland, grassland, meadows and hedge-banks. **Distribution** Most of northern Europe except Iceland and large areas of Scandinavia; sparse in Wales, Scotland and Ireland. **Flowers** June–September. **Image size** Two-thirds life-size.

This plant is very similar to lady's bedstraw (p.104) and other relatives, but is bigger and can be distinguished from its cousins by having smooth stems. The white flowers, 3mm long, appear at the top of square, branched stems. They are pollinated by flies, resulting in smooth, black fruits. The attractive ruffs of pale green foliage appear in whorls of between six and eight leaflets. Like its related species, hedge bedstraw was once used as a strewing herb. From the roots it is possible to make a red dye, while the stems and leaves were used for a yellow colouring. Bedstraw was occasionally added to flavour and colour cheese.

Feverfew [3]

Tanacetum parthenium (Asteraceae or Compositae)

Biennial or perennial. Height 60cm. **Habitat** Chalky or neutral soil on scrubland, rocky habitats and cultivated land near buildings. **Distribution** Northern Europe to southern Sweden. **Flowers** July–September. **In gardens** Well-drained, sandy soil, full sun. Edging for borders. **Image size** Two-thirds life-size.

Immediately identifiable by its pungent aroma, feverfew produces daisy-like 10-25mm flowers with a flat, bright yellow disc surrounded by a collar of blunt-tipped white rays. The yellowish-green leaves are pinnate with deeply divided, downy leaflets on hairy stems. Bees dislike the pungent aroma – not surprisingly, as the leaves contain volatile oils with insecticidal properties. Taken internally, feverfew is an old remedy for migraines; it was also used topically to soothe insect bites and bruising.

Lesser stitchwort [4]

Stellaria graminea (Caryophyllaceae)

Perennial. Height 15–60cm. **Habitat** Chalky soils in woodland, grassland and hedgerows. **Distribution** All northern Europe except Faeroe Is. **Flowers** May–August. **Image size** Two-thirds life-size.

The lesser stitchwort is easily recognised by its straggly form and grass-like foliage. The white starry flowers, 5–12mm across, have petals cut more deeply than those of the greater stitchwort (p.32) and are pollinated mainly by flies; the leaves are narrower. Unlike the greater stitchwort, this species is not credited with any medicinal properties.

Greater butterfly orchid [1]

Platanthera chlorantha (Orchidaceae)

Tuberous perennial. Height Up to 60cm. **Habitat** Woods, scrub and open grassland. **Distribution** All northern Europe. **Flowers** June–July. **Image size** Half life-size.

The statuesque greater butterfly orchid has elegant spikes of greenish-white flowers which look vaguely like alighting butterflies. They have six petals; the outer three spread, while the upper two of the inner whorl form a hood. The lower petal forms a long lip with an exaggerated down-curving spur. Only long-tongued butterflies and moths can get at the nectar in the spur, and the flowers are mostly visited at night by pollinating moths attracted by the delicate vanilla scent. The oval leaves are deeply veined and grow at the base of the flower spike. This species is easily confused with the lesser butterfly orchid (*Platanthera bifolia*), which is smaller in all its parts and has parallel rather than converging pollinia.

Sweet Alison, alyssum [2]

Lobularia maritima (syn. *Alyssum maritimum*; Brassicaceae or Cruciferae)

Annual or short-lived perennial. Height Up to 30cm. **Habitat** Sand or shingle. **Distribution** Introduced throughout northern Europe. **Flowers** June–September. **In gardens** Well-drained soil in full sun. Summer bedding and coastal gardens. **Image size** One-third life-size.

For anyone used to seeing sweet Alison penned in as an edging plant in bedding schemes, coming across this sprawling plant in the wild will be a surprise. Sweet Alison is particularly at home on shingle beaches where conditions are as near as possible to its native habitats in southern Europe. The tiny, pure white flowers are crammed together in dense rounded heads. Each has four petals, and they exude a sweet scent that attracts pollinating insects. The slen-der, pointed foliage and stems are silvery-grey and covered in soft hairs, enabling them to tolerate drying coastal conditions.

Pale persicaria [3]

Polygonum lapathifolia (Polygonaceae)

Annual. Height Up to 1m. **Habitat** Bare and waste land, and pond and river margins. **Distribution** Throughout northern Europe. **Flowers** June–October. **Image size** Half life-size.

Pale persicaria is either a natural marginal plant of ponds and rivers or a 'weed' of wasteland, depending on your perspective. It is a member of the dock family and produces stout stems, which may be decorated with red spots. The oval or linear leaves have yellow glands on their undersides and are arranged alternately up the stem. The spikes of tiny white flowers emerge at the leaf axils; the petals and stalks also have yellow glands. Pale persicaria is often found growing alongside redshank (*Persicaria maculosa*; syn. *Polygonum persicaria*), and the two are often confused. Redshank, however, has pale to deep pink flowers and no glands on the flowers, flower-stalks or leaves.

Watercress [4]

Rorippa nasturtium-aquaticum (syn. *Nasturtium officinale*; Brassicaceae or Cruciferae)

Aquatic perennial. Height Stems up to 1m long. **Habitat** Freshwater streams and rivers, ponds, ditches and muddy places. **Distribution** All northern Europe except Faeroe Is and Iceland. **Flowers** May–October. **Image size** Twice life-size.

The peppery flavour and nutritional value of watercress leaves ensure that it is as popular eaten raw or cooked. But it may harbour liver-fluke larvae, which are carried by water snails that feed on the leaves. To be safe, pick wild leaves only from plants growing in clear, fast-flowing water – not from ponds, ditches or banks – and rinse thoroughly or make into watercress soup.

The evergreen foliage is made up of glossy pinnate leaves with rounded leaflets. They are topped by racemes of small pure white flowers. The older leaves have more flavour than the young green foliage.

Common milkwort [5]

Polygala vulgaris (Polygalaceae)

Perennial. Height 30cm. **Habitat** Grassland, heathland and dunes. **Distribution** All northern Europe. **Flowers** May-September. **Image size** One-third life-size.

Common the delightful milkwort may be, but it often goes unnoticed. The flowers, 5–8mm long, have an unusual structure: the tube-like petals, which are long and narrow, are partly concealed by two large but shorter sepals, which may be white, blue, purple or pink. The flowers grow in loose spikes; once they are fertilised and the fruits set, the sepals turn green. The leathery leaves are oval at the base of the stem but more pointed and spear-shaped farther up. The leaves of heath milkwort (*P. serpyllifolia*) are arranged in opposite pairs, instead of alternately. Milkwort was traditionally used to encourage the flow of a mother's milk.

Sea kale [6]

Crambe maritima (Brassicaceae or Cruciferae)

Perennial. Height 40–60cm. **Habitat** Shingle and sand beaches. **Distribution** All northern Europe. **Flowers** June–August. **In gardens** Fertile, well-drained soil, full sun, shelter from winds. Herbaceous borders, coastal, wild-flower and woodland gardens. **Image size** Half life-size.

Before sea kale became fashionable in the late 18th century, clumps of its curly blue-green foliage and clusters of 10–15mm white flowers were quite a common sight along coasts. Unfortunately it became a victim of its own culinary success, and its numbers have generally dwindled in the wild. But sea kale is now available from nurseries, which sell it as a plant suitable for both wild-flower and coastal gardens. The sturdy succulent leaves, with their waxy coating, easily cope with desiccating sea winds. Gently boil the young white stems and smother in butter, like asparagus. But don't eat the leaves; they are, by all accounts, tasty but as tough as old boots no matter how long you boil them.

Burnet rose [1]

Rosa pimpinellifolia (Rosaceae)

Perennial shrub. Height Up to 1m. **Habitat** Coastal areas and dry sandy soil elsewhere. **Distribution** All northern Europe except Finland and Faeroe Is. **Flowers** May–July. **In gardens** Humus-rich, well-drained soil in full sun. Wild-flower gardens and informal shrubberies. **Image size** Two-thirds life-size.

Anyone walking unwittingly into a clump of burnet rose will quickly learn why it was originally called *spinosissima*, or 'most spiny'. It is happiest creeping through dunes, where its suckering habit forms large patches. Each leaf has up to nine toothed leaflets, rather like those of burnet saxifrage (p.67) – hence both the Latin and common names. The cupped flowers are usually creamy-white but may be pink. The hips are round and berry-like, with the five long sepals still attached; they are a dark enough purple to appear almost black. You can plant the burnet rose in the garden – but beware its vigorous spreading nature and savage spines.

Dropwort [2]

Filipendula vulgaris (Rosaceae)

Perennial. Height 60cm. **Habitat** Dry grassland, roadsides and woodland edges. **Distribution** All northern Europe. **Flowers** June–September. **In gardens** Damp soil, partial shade. Bog gardens, pond margins. **Image size** Two-thirds life-size.

Although closely related to meadowsweet (p.71), there are three features that make dropwort easy to distinguish. The creamy-white flowers, 8–16mm across – which, like those of its cousin, have six petals and prominent stamens – are larger and fewer. On the other hand, it has smaller and more numerous leaflets; and instead of twisted clusters of fruits, the dropwort produces straight clusters. It has made the transition from wild flower to garden ornamental with ease and there are several attractive cultivars, such as the double-flowered white 'Flore Pleno' and the rose-pink 'Rosea'.

White melilot [3]

Melilotus alba (Fabaceae or Leguminosae)

Annual. Height 60–120cm. **Habitat** Fields, wasteland and roadsides. **Distribution** Holland and Scandinavia; naturalised in Belgium, Britain and Ireland. **Flowers** July–September. **Image size** Two-thirds life-size.

Lupin-like, white spires of 4–5mm flowers on tall, upright stems are each accompanied by three oval leaflets with sharply toothed edges. The long upper petal distinguishes white melilot from closely related species. The oval seedpods are brown and hairless. Like all its relations, white melilot is sweetly scented; when dry, it smells strongly of new-mown hay. The flowers brim with nectar, luring bees and hoverflies, which press down on the lower petals as they land and brush on the stigma, depositing pollen collected from another plant. In France it is associated with St John and ceremonies celebrating St John's Eve.

Enchanter's nightshade [4]

Circaea lutetiana (Onagraceae)

Perennial. Height 20–70cm. **Habitat** Damp woodland and hedge-banks. **Distribution** All northern Europe except far north. **Flowers** June–August. **Image size** Two-thirds life-size.

This plant is steeped in superstition. It was said to give protection from elvish spells in Anglo-Saxon England. In Germany, it was associated with witchcraft. French names include *herbe à la magicienne* and *herbe aux sorcières*. The Latin name comes from the mythical Greek sorceress Circe.

It is an elegant plant with downy stems and paired oval leaves. The flowers are white, with two deeply notched petals, in slender, loose spikes. The drooping fruits are covered with hooked bristles to latch on to passing animals – unlike all the related willowherbs, whose seed is distributed by wind. The plant's creeping rhizome is brittle and can regenerate from the smallest piece, so it spreads easily in gardens.

Nottingham catchfly [5]

Silene nutans (Caryophyllaceae)

Perennial. Height Up to 80cm. **Habitat** Chalky soil on disturbed ground, rocky places, grassland and shingle. **Distribution** Northern Europe except the northernmost tip and Ireland; rare in Britain. **Flowers** May–August. **Image size** Two-thirds life-size.

The Nottingham catchfly's sweet-smelling flowers open fully only at night. The deeply cut petals, rolled up by day, then unfurl to form large spidery flowers with prominent stamens. They have evolved a fascinating technique for cross-pollination. Each flower lasts three nights and the scent attracts night-flying moths. On the first and second nights, the flower puts out five stamens to discharge their pollen. On the third, the styles take the place of the stamens, so the stigmas – which by now have had plenty of time to mature – can receive pollen from a younger flower on the same plant.

Rough chervil [6]

Chaerophyllum temulum (syn. *C. temulentum*; Apiaceae or Umbelliferae)

Biennial. Height Up to 1m. **Habitat** Hedges, wood margins and grassland. **Distribution** All northern Europe except Finland, Iceland and Faeroe Is. **Flowers** June–July. **Image size** Two-thirds life-size.

Once cow parsley (p.32) has finished flowering in June, this species is the next wayside 'parsley'. It can be distinguished from its predecessor by the stems, which are covered in bristly hairs, are either purple-spotted or completely purple, and swollen under each leaf node. The leaves are divided into lobes, with each lobe further divided; they have downy hairs on both sides. The umbrella-shaped flower heads have rays of varied lengths ending in small umbels of white flowers. Each has five petals, with a notch at the tip. Unlike cow parsley, rough chervil can be toxic. It is not as poisonous as some of its cousins, but consuming it is similar to being drunk.

Sulphur clover [1]

Trifolium ochroleucon (Fabaceae or Leguminosae)

Perennial. Height 50cm. **Habitat** Grassland on heavy clay soil. **Distribution** Eastern England **Flowers** June–July. **Image size** Two-thirds life-size.

The most likely place for spotting the egg-shaped flower heads of the rare sulphur clover is in the east of England. Each flower head, on closer inspection, consists of a mass of tiny pale yellow flowers, cupped in a pair of leaves; each leaf consists of three oval leaflets. Once the flowers have been fertilised they turn pale brown.

Wild carrot [2]

Daucus carota (Apiaceae or Umbelliferae)

Annual or biennial. Height 30–100cm. **Habitat** Grassland. **Distribution** Throughout northern Europe. **Flowers** June–August. **Image size** Two-thirds life-size.

An extremely variable plant, the wild carrot can be hairy or hairless, short or tall. At the centre of dense, concave umbels of white 2mm flowers, there is usually a single purple or dark red flower. This characteristic distinguishes the plant from its cousins in the same family. As the seeds ripen, the umbels contract to form a cup-like shape, resulting in the common country name 'bird's nest'. In the 16th and 17th centuries, the feathery leaves, which have a faint carroty scent, were used to decorate ladies' head-dresses. Unlike our culinary carrot, the thick taproot is white and woody, rather than orange and fleshy.

The seeds of wild carrot were thought to relieve flatulence and coughs. Today, as a rich source of vitamin A, the oil is an ingredient of anti-wrinkle creams.

White stonecrop [3]

Sedum album (Crassulaceae)

Perennial. Height 7–15cm. **Habitat** Rocky places and roadsides. **Distribution** Throughout northern Europe except Faeroe Is and Iceland. **Flowers** June–August. **Image size** Two-thirds life-size.

Straggling stems of white stonecrop form low-growing mats of alternate, succulent foliage; flat clusters of white, starry flowers, 6–9mm across, appear at the stem tips. The whole plant appears to recline on the bare rock that is its natural habitat. This, according to the 19th-century garden writer J.W. Loudon (with tongue firmly in cheek), could have explained the source of the name, *sedere* meaning 'to sit'. The up-to-date (and more likely) version of the story, however, is that *Sedum*, the Latin name for the houseleek, comes from a word meaning 'to assuage'. This refers to the healing properties of the related small houseleek, which was known to the 17th-century herbalist Nicholas Culpeper. Its uses included making a cooling ointment for piles.

Hemlock water dropwort [4]

Oenanthe crocata (Apiaceae or Umbelliferae)

Perennial. Height 30–90cm. **Habitat** Marshes, margins of ponds and other wet habitats, grassland and wood edges. **Distribution** Belgium, Britain, Ireland and France. **Flowers** June–August. **Image size** Two-thirds life-size.

The hemlock water dropwort has probably been responsible for more human deaths than any other plant growing in its range – often among people who mistake it for wild celery (p71). Despite a deceptively sweet, wine-like scent and flavour, the entire plant is extremely poisonous. Yet it is very handsome, with sturdy, grooved stems and pinnate leaves of rounded celery-like leaflets. On close inspection, the large flower heads are seen to be made up of a number of umbels, each consisting of hundreds of tiny white flowers. Each individual flower is only 2mm across.

Hare's-foot clover [5]

Trifolium arvense (Fabaceae or Leguminosae)

Annual or biennial. Height 20cm. **Habitat** Sandy soils in dry grassland, heaths, verges and wasteland. **Distribution** All northern Europe except the northernmost tip, Faeroe Is, Iceland and northern Scotland. **Flowers** June–September. **Image size** Two-thirds life-size.

The obvious similarity of this plant's flowers to a hare's foot has been recorded since the 16th century. The silky, 25mm heads of white or pinkish flowers – each about 4mm long – are egg-shaped. When they dry out and turn brown they resemble their namesake even more closely. The slender trifoliate leaves – each leaflet much narrower than those of most other clovers – and upright stems are also covered in a soft layer of whitish hairs. Unlike in parts of mainland Europe, hare's-foot clover is now rarely seen on arable land in Britain, as a result of intensive farming practices.

Cleavers, goosegrass [1]

Galium aparine (Rubiaceae)

Annual. Height 1.2m. **Habitat** Woodland and scrub. **Distribution** All northern Europe except the northernmost tip. **Flowers** May–September. **Image size** Two-thirds life-size.

Cleavers deserves top marks for tenacity. With leaves, stems and fruits smothered in backward-facing bristles, it maximises every opportunity to survive and distribute its seed. It clings to everything it touches – hence the name – helping it to clamber up through dense plants towards daylight. The insignificant white flowers are followed by hooked, round fruits, or burrs, which cling tenaciously to animal fur and people's clothing. Leaves appear in whorls of up to seven at intervals along the stems.

Cleavers has been used variously to make a tea to cure insomnia and an infusion for sunburn. In Sweden the stems were woven to make a rough filter to strain milk, and in France the crushed plant was made into a poultice to soothe sores and blisters. Despite the rough texture, geese love to eat it.

White bryony [2]

Bryonia dioica (Cucurbitaceae)

Climbing perennial. Height 4m. **Habitat** Hedgerows and woodland edges. **Distribution** Belgium, southern Britain, France, Holland and Germany. Introduced in Ireland. **Flowers** May–September. **Image size** Two-thirds life-size.

White bryony is reminiscent of a cucumber plant, to whose family it belongs. Five-lobed palmate leaves and clusters of greenish-white 10–18mm flowers are borne on vigorous, winding stems whose tendrils grasp hedgerows and other supports. *Dioica* ('two houses') refers to the male and female flowers, which are on separate plants. But the non-cucumber-like green berries ripen bright red and are poisonous – as are the roots, which used to be passed off as mandrake. There are tales of cattle poisoning, but in France it was used to reduce the flow of a mother's milk.

Gipsywort [3]

Lycopus europaeus (Lamiaceae or Labiatae)

Perennial. Height 30–100cm. **Habitat** Wet woodlands, marshes, streams and ditches. **Distribution** Throughout northern Europe except the northernmost tip. **Flowers** July–September. **Image size** Two-thirds life-size.

Whorls of small (4mm) white flowers nestle at the nodes of distinctive, toothed leaves all along stiff, square stems – which are typical of the mint family. The leaves are arranged in opposite pairs, distinguishing gipsywort from related plants. The plant was used to make a strong black dye and got its common name from an old belief that gipsy people used the dye to darken their skin. Unlike many other members of the mint family, gipsywort has no scent. There are no records of any culinary uses and only vague references in herbals to it being used as an astringent and sedative.

Yarrow [4]

Achillea millefolium (Asteraceae or Compositae)

Perennial. Height 80cm. **Habitat** Chalky or slightly acid soil in grassy places. **Distribution** All northern Europe. **Flowers** July–October. **In gardens** A common lawn weed, but are cultivars grown in borders; any soil, full sun and used for cutting. **Image size** Two-thirds life-size.

This tall, elegant herb is the stuff of magic and mythology. Its healing properties are legendary, and it was named after Achilles, who used it to mend his soldiers in the Battle of Troy. Druids used yarrow to divine the weather, and in Ireland bunches of it were thought to keep illness at bay. It also confers disease-resistance on nearby plants.

The feathery, fern-like leaves consist of hundreds of fine leaflets – hence the name *millefolium* ('thousand-leaf'). Tiny white daisy-like flowers, with a yellow or cream disc, grow in flat heads up to 12cm wide.

Yarrow is a useful ornamental, especially for coastal gardens, where the tiny leaflets reduce moisture loss and the supple stems tolerate buffeting winds.

Marsh bedstraw [5]

Galium palustre (Rubiaceae)

Perennial. Height Up to 1m. **Habitat** Damp or wet places. **Distribution** All northern Europe. **Flowers** June–August. **Image size** Two thirds life-size.

The marsh bedstraw's thin creeping stems allow it to weave its way nimbly through taller, supporting plants. The clusters of white flowers, like miniature 2–3mm stars, have red anthers and are pollinated by bees. They appear on stems that emerge from whorls of five or six polished leaves, which have backward-facing prickles on their edges. The leaves turn black as they dry out – a useful method of distinguishing this species from the rather similar fen bedstraw (p.60).

Catmint [6]

Nepeta cataria (Lamiaceae or Labiatae)

Perennial. Height 30–100cm. **Habitat** Mostly chalky soil in hedgerows, roadsides and rocky habitats. **Distribution** Belgium, Britain, France and Holland; naturalised in Germany, Ireland and Scandinavia. **Flowers** June–September. **In gardens** Any well-drained soil in full sun. Wild-flower gardens, herb gardens and borders. **Image size** Two-thirds life-size.

The true catmint produces upright spikes of densely packed white flowers, 7–10mm long, with dark purple spots to guide pollinating bees to a rich source of nectar. The leaves look rather nettle-like, and it is their minty scent that is so attractive to cats. It is no longer used as a medicinal herb to the extent that it once was, but that is not to underestimate its properties – which are thought to be particularly effective against childhood illnesses such as flu and measles. It is sometimes grown in gardens, but much more common is a hybrid, *N. x faasenii*, and its cultivars.

Hemlock [1]

Conium maculatum (Apiaceae or Umbelliferae)

Annual or biennial. Height 1.5–3m. **Habitat** Damp places, roadsides and wasteland. **Distribution** All northern Europe except Faeroe Is and Iceland. **Flowers** June–July. **Image size** Half life-size.

Infamously poisonous, hemlock can easily be mistaken by the untrained eye for several other members of the parsley family. It has the typical umbels of tiny white flowers and finely divided leaves. But the hollow stems have bold purple blotches – a clear warning signal. This is reinforced by the noxious, mousy smell, which fades when the plant dries out. There have been many cases of poisoning when the leaves were mistaken for parsley, and among children who made whistles from the hollow stems.

Hemlock root was added by Shakespeare's three witches, in *Macbeth*, to their bubbling poisonous brew. It is thought to have been responsible for the lingering execution of the Greek philosopher Socrates.

Carline thistle [2]

Carlina vulgaris (Asteraceae or Compositae)

Biennial. Height Up to 60cm. **Habitat** Dry grassland and dunes, on chalky or sandy soil. **Distribution** All northern Europe except far north. **Flowers** July–September. **Image size** Half life-size.

Carline thistle produces distinctive 2–4cm flower heads surrounded by spiny leaf-like bracts. What look like pale yellow-white ray florets are actually inner bracts that contrast nicely with the brownish tubular disc florets in the centre – each head like a miniature sunburst. They open in dry weather and close when it is damp, acting like floral barometers. The flower heads appear at the tips of the branched stem and the lobed leaves are armed with many prickles. Carline thistle was once used as an antiseptic. King Charlemagne used the plant – supposedly on the advice of an angel – to cure his army of plague – hence the name 'carline'.

Greater burnet-saxifrage [3]

Pimpinella major (Apiaceae or Umbelliferae)

Perennial. Height 1.2m. **Habitat** Dry grassy areas and shady woodland edges. **Distribution** All northern Europe except the northernmost tip. **Flowers** June–August. **In gardens** Moist soil, full sun, partial shade. Wild-flower gardens. **Image size** Half life-size.

The greater burnet-saxifrage is neither a burnet nor a saxifrage. It is one of many umbelliferous flat-topped perennials of the parsley family. This species is a variable plant, producing 3mm flowers in many shades, from white to dark pink, and deeply cut leaves larger than those of the closely related burnet saxifrage (p.67). Lesser water parsnip (*Berula erecta*) has a similar appearance, but is found growing in shallow fresh water.

The French name focuses not on its pleasant appearance but on the foul goat-like smell of the roots; it is called *boucage*, from *bouc* (billy goat). Putrid roots aside, the garden variety *P. major* 'Rosea' has attractive deep rose-pink umbels.

Thyme-leaved sandwort [4]

Arenaria serpyllifolia (Caryophyllaceae)

Annual. Height Up to 30cm. **Habitat** Chalk downs, arable land, bare sandy places and walls. **Distribution** All northern Europe except far north. **Flowers** April–September. **Image size** One-third life-size.

Sprawling, wiry stems and tough thyme-like leaves protect this species of sandwort from marauding rabbits, who find it so unpalatable that they leave it in peace. Otherwise there is little that is very distinctive about it. The clusters of tiny white star-like flowers with green sepals are similar to those of common chickweed (p.75) – although, unlike chickweeds, the petals are undivided and are shorter than the sepals. They are followed by tiny kidney-shaped nutlets. This is one plant that would benefit from close inspection with a magnifying glass; if the flowers were 2cm across, every gardener would want one.

Fairy flax [5]

Linum catharticum (Linaceae)

Annual. Height Up to 20cm. **Habitat** Well-drained heaths, moors, grassland and rocks. **Distribution** Throughout northern Europe. **Flowers** May–September. **Image size** Life-size.

Typically, flax is large and blue-flowered, but fairy flax is very untypical. Yet with its narrow opposite leaves and loose flower heads, it looks like a scaled-down version of the larger flaxes – except for the white flowers. They are 4–6mm across, have five petals and slender stalks.

Fairy flax gained its name in the Middle Ages, when anything of the fairy world was believed to be powerful, mischievous, even malevolent. It is also called purging flax – the Latin *catharticum* means exactly that. The plant was infused, either in wine or water, and produced an extreme reaction – so much so that it was effective against intestinal worms.

Cloudberry [6]

Rubus chamaemorus (Rosaceae)

Creeping perennial. Height Up to 20cm. **Habitat** Mountain bogs and moors. **Distribution** Britain, Ireland, Germany and Scandinavia. **Flowers** June–August. **In gardens** Moist acid soil in sun. Acid borders. **Image size** Half life-size.

Cloudberry is a distinctive plant among *Rubus* species – the genus that includes brambles and the raspberry. It is low-growing and has roughly hairy (rather than prickly) stems and attractive lobed foliage, with a crinkled texture. The white male and female five-petalled 15–20mm flowers appear on separate plants and are followed by sumptuous berries, which ripen orange or red. Despite the romantic notion that the cloudberry gets its name by growing on the peaks of cloudy mountains, the name is actually derived from the Old English *clud*, which means 'rock' or 'hill' and accurately describes the habitat where it thrives.

Canadian fleabane [1]

Conyza canadensis (Asteraceae or Compositae)

Annual. Height 1.2m. **Habitat** Wasteland, cultivated land, dunes and walls. **Distribution** Naturalised throughout northern Europe except Ireland. **Flowers** July–September. **Image size** Two-thirds life-size.

A North American native, Canadian fleabane was first recorded in Europe in the mid-17th century at the Botanic Garden at Blois, in northern France. It reached England around 1690, as seed allegedly via a stuffed bird imported from America. The daisy-like flowers, 2–5mm across, appear in branched clusters on upright, hairy stems along with linear, alternate leaves covered in fine hairs. The downy, shuttlecock-like seeds are dispersed on air currents – the plant now grows all over northern Europe.

There are conflicting stories to explain the name 'fleabane'. Garden writer J.W. Loudon said it was thought to repel fleas and gnats, but herbalist Nicholas Culpeper was convinced that it was named thus because 'the seeds are so like fleas'.

Eyebright [2]

Euphrasia agg. (Scrophulariaceae)

Annual. Height and spread Variable, 5–35cm. **Habitat** Grassy habitats. **Distribution** All northern Europe except far north. **Flowers** May–October. **Image size** Two-thirds life-size.

In the Middle Ages the purple and yellow blotches on the petals of this plant were thought to resemble a bruised eye. As a result, eyebright was bestowed with not only its common name but all sorts of ophthalmic properties too.

There are more than 25 species of *Euphrasia* in the British Isles alone, but without specialist knowledge they are almost impossible to distinguish from one another. (Botanists term such a group an 'aggregate'.) Typically, however, they produce small, white flowers with purple veins, a yellow throat and a lower three-lobed lip.

The oval, toothed leaves are arranged in opposite pairs. Eyebrights are semi-parasitic plants, plying the roots of red clover, plantains and grasses for vital nutrients.

Chamomile [3]

Chamaemelum nobile (Asteraceae or Compositae)

Perennial. Height 15cm. **Habitat** Grassy habitats and roadsides on sandy soil. **Distribution** France and southern and eastern Britain; naturalised in Belgium and Germany. **Flowers** June–August. **In gardens** Free-draining soil in warm, sunny spots. Rock gardens, paving and lawns. **Image size** Two-thirds life-size.

Chamomile (or camomile) is best known as a garden plant. It forms creeping mats of fine, feathery foliage, which exude a strong apple-like scent when crushed – hence the name, which comes from the Greek *chamaimelon* ('apple on the ground'). Solitary flowers, 18–25mm across, of yellow discs and white rays, emerge at the tip of each spindly stem.

Grown as a lawn, it needs plenty of maintenance. The flower heads should be cut down regularly, and the only way to keep it free of weeds is by hand. 'Treneague' is a non-flowering cultivar useful for lawns, or grow the flowering form between paving slabs.

The powerful scent makes chamomile a good companion plant for onions, keeping away flying pests, and it is known as 'the physician's plant' because it can revive its ailing neighbours. Stinking camomile (*Anthemis cotula*) is a similar plant, but with a strong smell and a reputation for causing blistered hands at harvest time.

Fen bedstraw [4]

Galium uliginosum (Rubiaceae)

Perennial. Height 10–50cm. **Habitat** Wet places. **Distribution** All northern Europe. **Flowers** June–August. **Image size** Two-thirds life-size.

A more bristly species than its close relative marsh bedstraw (p.56), the hay-scented fen bedstraw has spindly stems tipped with clusters of 2.5–3mm white flowers. It is easily confused with the marsh bedstraw but can be distinguished by its whorls of six to ten leaves, which have bristly edges. Also, unlike its cousin, the plant does not turn black when it dries out.

Wavy bittercress [5]

Cardamine flexuosa (Brassicaceae or Cruciferae)

Biennial or perennial. Height 50cm. **Habitat** Damp shade. **Distribution** All northern Europe. **Flowers** April–September. **In gardens** A common weed. **Image size** Two-thirds life-size.

This plant is very similar to its annual cousin the hairy bittercress (*Cardamine hirsuta*), except that it is taller. If you want to be sure that a plant is the wavy kind, count the stamens in the white 3–4mm flowers; *C. flexuosa* has six, *C. hirsuta* only four. Like several other species of the cabbage and mustard family, the leaves contain the tangy oil that gives bittercress its name. The wavy bittercress distributes seeds via exploding pods, so it is extremely successful at colonising patches of bare soil. If it encroaches on your garden, it is easy to pull out by hand when the soil is damp; failing that, lever it out with a hoe.

Bog stitchwort [6]

Stellaria uliginosa (Caryophyllaceae)

Perennial. Height 15–60cm. **Habitat** Acidic soil in damp habitats; woodland rides. **Distribution** Most of northern Europe except far north. **Flowers** May–June. **Image size** Two-thirds life-size.

Stellaria is the Latin for 'star' and an apt name for the tiny (5–7mm) white flowers produced by the bog stitchwort. The starry effect is accentuated by the sepals, which are longer than the petals. The blue-grey, lanceolate leaves are arranged in pairs along creeping stems.

Water chickweed [1]

Myosoton aquaticum (Caryophyllaceae)

Trailing perennial. Stem length Up to 1m.
Habitat Marshy areas; banks of streams and rivers.
Distribution Throughout most of northern Europe except
Faeroe Is, Iceland and Ireland. **Flowers** June–August.
Image size Half life-size.

A straggling plant with long, weak stems,
water chickweed makes as much use as it
can of surrounding vegetation to scramble
towards the light. The leaves are opposite,
with a heart-shaped base and pointed tip;
they are thin and often have a wavy edge.
The flowers are held in a loose cluster. Each
has five narrow, blunt sepals, which have a
wide white margin. The petals are white,
longer than the sepals, and cut almost to
their base. In the British Isles it is found
mainly in England, and is only introduced
(not native) to Scotland. It seems to avoid
the extremes of both alkaline chalk and
limestone areas, and the acid soils over
granite rocks – a very British compromise.

Common hemp nettle [2]

Galeopsis tetrahit (Lamiaceae or Labiatae)

Annual. Height Up to 50cm. **Habitat** Arable land,
clearings in woods and other damp areas.
Distribution Throughout Northern Europe.
Flowers July–September. **Image size** Life-size.

The common name comes from the vague
similarity of the oval leaves to those of the
unrelated *Cannabis*; they are narrower than
those of dead nettles, which are related.
It has hairy, usually branched stems. The
flowers are in whorls at the leaf nodes at the
top of stems. They are white and two-
lipped; the upper lip forms a hood, while
the lower has three lobes. The central lobe
usually has dark markings. Common hemp
nettle is a very variable species.

White horehound [3]

Marrubium vulgare (Lamiaceae or Labiatae)

Perennial. Height 60cm. **Habitat** Grassy habitats and
wasteland. **Distribution** Scattered through northern
Europe, including south and west coasts of Britain;
rare everywhere. **Flowers** June–September. **In gardens**
Dry, well-drained soil, full sun. Mixed borders, containers.
Image size Twice life-size.

Despite having been used medicinally since
Roman times – and still an ingredient of
herbal cough syrups and pastilles – wild
white horehound (or hoarhound) is now
rare in northern Europe. The square stems
and wrinkly leaves are covered in downy
hairs – probably the origin of the common
name, from the Anglo-Saxon *har* (grey; the
same root as for 'hoar'). The silvery-grey
leaves smell like thyme. At each axil is a
cushion-like whorl of small white flowers,
12–15mm long. Infusing the leaves in
water makes an insecticide that has been
used against caterpillar pests, and is an
effective fly-killer if mixed with milk.
White horehound is easily cultivated in the
garden. Harvest the leaves just as the plant
comes into flower in spring.

Black nightshade [4]

Solanum nigrum (Solanaceae)

Annual. Height 70cm. **Habitat** Nutrient-rich soil on bare
or disturbed ground and wasteland. **Distribution** All
northern Europe except the northernmost tip. **Flowers**
July–October. **Image size** Life-size.

Black nightshade belongs to a huge family;
the genus *Solanum* alone contains at least
1,700 species, including the potato. Many
family members are poisonous; black nightshade is less so than some, but it does
contain the poisonous alkaloid solanine.
The plant is so called because its spreading
stems are blackish, as are the berries. The
10–14mm flowers that precede them are
white and starry, with reflexed petals and
prominent yellow anthers – very similar to
those of the potato plant.

Arrowhead [5]

Sagittaria sagittifolia (Alismataceae)

Aquatic perennial. Height Up to 1m. **Habitat** Ponds,
canals and slow-flowing rivers. **Distribution** All northern
Europe except the far north. **Flowers** July–August.
In gardens Full sun. Marginal plant for informal pools.
Image size One-quarter life-size.

This handsome aquatic plant makes a
striking impression, with its deeply veined
arrow-shaped leaves held well above the
water's surface. The fat buds open to reveal
three white petals, tinged purple at their
base; the flowers are 20–25mm across.
There are separate male and female flowers
on the same plant. The females huddle in
whorls on the lower half of the fleshy,
triangular stem. The male blooms, which
have a mass of purplish-brown stamens
with arrow-shaped anthers, grow on long
flower stems farther up. The cultivar
'Flore Pleno' has double flowers. Although
arrowhead is edible, few books mention any
culinary uses.

Common water plantain [6]

Alisma plantago-aquatica (Alismataceae)

Semi-aquatic perennial. Height Up to 1m. **Habitat**
Pond margins, canals, rivers and marshes. **Distribution**
All northern Europe except far north. **Flowers**
June–August. **In gardens** Containers of ordinary soil.
Pool margins. **Image size** One-sixth life-size.

The common water plantain, one of several
attractive aquatic plants in its family,
produces sprays of strongly veined, oval
leaves on long fleshy stems. Handsome
branched stems of white flowers, sometimes
with a purplish tinge, emerge from the
centre of the clump of foliage. They unfurl
only in the afternoon for a few hours before
closing again at dusk. All the flower parts –
petals, sepals and stamens – are in threes;
they are pollinated by flies attracted by
droplets of nectar. The leaves were thought
to relieve dropsy and in North America
were used to soothe rattlesnake bites.

Scentless mayweed [1]

Matricaria perforata (syns. *M. inodora*,
Tripleurospermum inodorum; Asteraceae or Compositae)

Biennial or perennial. Height 60cm. **Habitat**
Roadsides, wasteland and cultivated land.
Distribution Throughout northern Europe. **Flowers**
July–September. **Image size** Half life-size.

This is the most common of all the may-
weeds. The common name comes from the
Latin *matrix* (womb), suggesting that it was
used medicinally for female disorders. It has
nothing to do with the month of May – it
flowers from midsummer to early autumn
– but it *is* scentless. The solitary, daisy-like
15–50mm flower heads on erect stems,
have a yellow disc surrounded by a collar of
pure white rays. The foliage is finely divided.

Hedge bindweed, bellbine [2]

Calystegia sepium (Convolvulaceae)

Climbing perennial. Height 3m. **Habitat** Hedge-banks,
wasteland, woodland and river margins. **Distribution**
Throughout northern Europe except the far north.
Flowers June–September. **In gardens** Usually regarded
as a weed. **Image size** Half life-size.

The long, pointed buds of the hedge
bindweed unfurl into glorious pure white
trumpets, 30–50mm across, rightly earning
it the title of morning glory. When the moon
is very bright, the flowers stay open at night
and attract convolvulus hawk moths, which
pollinate the flowers while lapping up the
nectar with their long tongues.

The charming blooms are supported by
spindly stems, which twine anti-clockwise
and are much stronger than they seem. The
bright green arrow-shaped leaves are arranged
alternately along the stems; they are one of
the features that distinguish hedge bindweed
from its pink-flowered cousin field
bindweed (p.132), whose leaves are more
shield-shaped. Bindweeds are generally
regarded as pests, particularly on arable
land. But the roots of this species are less
tenacious than those of its relatives, so it
has a better reputation as an attractive screen.

Rue-leaved saxifrage [3]

Saxifraga tridactylites (Saxifragaceae)

Annual. Height Up to 10cm. **Habitat** Dry, bare ground,
sandy heaths, walls and rocks. **Distribution** All northern
Europe except far north. **Flowers** June–September.
Image size Life-size.

The small white flowers of rue-leaved
saxifrage contrast beautifully with the
lobed, slightly fleshy foliage, which has a
soft reddish tinge and is covered in sticky
hairs. This red tint helps to protect the
plant's green chlorophyll from light – the
more direct the light, the redder the leaves
become. The flowers grow in loose clusters
on long stalks, and have five rounded petals.

Starry saxifrage [4]

Saxifraga stellaris (Saxifragaceae)

Perennial. Height 8–20cm. **Habitat** Damp mountain
habitats, wet rocks and ledges, stream-banks and
marshes. **Distribution** All northern Europe except
southern Britain, Belgium, Holland and Denmark.
Flowers June–August. **In gardens** Moist yet
well-drained soil in sun. Rock gardens. **Image size**
Half life-size.

It is astonishing that some of the most
fragile-looking wild flowers are among the
toughest, able to survive even the most
hostile environments. Starry saxifrage is no
exception. As the name suggests, it
produces pretty star-shaped flowers,
10–15mm across. They have oval white
petals with two yellow spots at the base and
contrasting pinkish anthers. The flowers
grow on wiry branched stems that rise out
of a rosette of toothed oval leaves, whose
upper surface is covered in hairs.

The whole plant is firmly anchored to the
wet rocks where it flourishes by roots that
can penetrate every crack and fissure – the
genus name *Saxifraga* means 'rock-breaker'.
As a result of this supposed ability to
break stones, starry and other saxifrage
species were regularly used as a cure for
kidney and gall stones.

Heath bedstraw [5]

Galium saxatile (Rubiaceae)

Perennial. Height Up to 30cm. **Habitat** Dry grassland,
heaths, open woodland and rocky habitats.
Distribution All northern Europe. **Flowers** June–August.
Image size Life-size.

Heath bedstraw is an altogether daintier
plant than its relation lady's bedstraw
(p.104). Instead of dense panicles of yellow
flowers, it produces much looser panicles of
very sweet-scented white flowers at the tip
of each flower stem. The leaves are arranged
in whorls along the length of the stems;
they turn black when they dry out. Heath
bedstraw bears smooth rounded fruits,
unlike its relative, cleavers (p.56), whose
fruits are covered in tiny hooks.

Fenugreek, bird's-foot clover [6]

Trifolium ornithopodioides (Fabaceae or Leguminosae)

Annual. Height Up to 20cm. **Habitat** Open coastal
habitats on sandy soil. **Distribution** Britain, Ireland,
France, Holland and Germany. **Flowers** May–September.
Image size Life-size.

Instead of the rounded flower heads typical
of the clover family, fenugreek produces
small clusters of two to four white or
pinkish flowers at the tips of its sprawling
stems. They are followed by long seedpods –
the origin of the alternative common name.
The foliage is more characteristic of clovers,
consisting of trifoliate leaves that have oval,
finely toothed leaflets. This annual smells of
new-mown hay by virtue of the chemical
coumarin, which is also present in the
bedstraws (pp.42, 48, 58, 60, 64, 104).

The name fenugreek originally belonged
to a related Mediterranean and Asian plant,
Trigonella foenum-graecum, whose aromatic
seeds are used as a spice. The common
name is an Old English corruption of the
Greek, meaning 'Greek hay' and referring to
the scent. Confusingly, this name became
associated with the northern European plant
even though it has no culinary use.

Hogweed [1]

Heracleum sphondylium (Apiaceae or Umbelliferae)

Biennial or short-lived perennial. Height 60–180cm. **Habitat** Grassland, roadsides and open woodland. **Distribution** All northern Europe except the northernmost tip. **Flowers** April–September. **Image size** Two-thirds life-size.

This is the most common member of the parsley family. Like many of its relatives, it has white flowers, 5–10mm across, in flat-topped umbels and deeply lobed leaflets, all on hairy, upright stems. In July the umbels are a hotbed of activity for the orange soldier beetle, which uses them as a source of food and a comfortable bed for mating. The thick, hollow stems have long been used by children as water guns and peashooters. The plant used to be collected as fodder for pigs – hence the common name; the young shoots and leaves taste like asparagus when boiled.

Parsley water dropwort [2]

Oenanthe lachenalii (Apiaceae or Umbelliferae)

Perennial. Height 30–80cm. **Habitat** Damp grassland and marshland. **Distribution** Northern Europe to southern Scandinavia. **Flowers** June–September. **Image size** Two-thirds life-size.

The foliage is the feature that distinguishes the non-poisonous parsley water dropwort from the closely related but lethal hemlock water dropwort (p.55). The leaflets are grey-green and narrow rather than celery-like. Umbels with up to 15 rays of tiny white flowers, 2–3mm across, appear at the top of tall, straight stems. The name *Oenanthe* comes from the Greek words for 'wine' and 'flower'; it refers to the pleasant smell of the flowers, which is similar to those on grape vines.

Burnet saxifrage [3]

Pimpinella saxifraga (Apiaceae or Umbelliferae)

Perennial. Height 30cm–1m. **Habitat** Grassland and rocky habitats. **Distribution** Most of northern Europe. **Flowers** June–September. **Image size** Two-thirds life-size.

Like its larger relative the greater burnet saxifrage (p.59), this is neither a true burnet nor a saxifrage. But it could be regarded as a composite of two other plants from the same family. The flat-topped umbel of tiny (2mm) white flowers is similar to a wild carrot's (p.55), while the broad lower leaves look as if they should really belong to the salad burnet (p.164); the upper leaves are finer. The species name *saxifraga* refers to the plant's supposed ability to break up stones in the body. Chewing the peppery fresh root was thought to relieve toothache and in a decoction was meant to remove freckles. More pleasantly, bunches of burnet saxifrage add an aromatic flavour to beer and help to sweeten tart wines.

Bladder campion [4]

Silene vulgaris (Caryophyllaceae)

Perennial. Height 25–90cm. **Habitat** Dry, chalky soil in grassy habitats and wasteland. **Distribution** Throughout Britain except parts of north and west Scotland; northern Europe. **Flowers** May–September. **Image size** Two-thirds life-size.

The 16–18mm white flowers of bladder campion grow in cymes on tall, upright stems. They have deeply notched petals, and exude a clove-like fragrance at night. The plant's name comes from the sepals, which form a curious inflated calyx looking like a miniature balloon. The flowers are mostly bisexual and are adapted for insect pollination. However, their supply of nectar is buried so deep inside the flower that some bees are known to bite through the base of the flower to get at the nectar. The blue-green waxy leaves are arranged in pairs and were once highly rated as a boiled vegetable, with a fragrance of fresh peas.

White water lily [5]

Nymphaea alba (Nymphaeaceae)

Aquatic perennial. Spread Up to 1.7m. **Habitat** Nutrient-rich aquatic habitats: slow-moving streams and rivers, lakes and ponds. **Distribution** Throughout northern Europe except Faeroe Is and Iceland. **Flowers** June–September. **In gardens** Too vigorous for all but large pools. **Image size** Two-thirds life-size.

It is no surprise that the pure beauty of the white water lily has meant that it is now widely cultivated and grown in large ornamental ponds and lakes. The slightly fragrant white flowers, 10–20cm across, start off cup-shaped but later turn into floating stars. They have yellow stamens. The rounded lily pads are dark green with reddish undersides, and the entire plant is anchored to the muddy bottom by fleshy, underwater stems. Immortalised in poetry for its beauty, the white water lily was also thought (questionably) to have medicinal properties, such as curing baldness. Hopeful Elizabethans believed – possibly inspired by the pure white blooms – that, in powder form, the rhizomes would encourage chastity. More prosaically, the underwater stems are sometimes eaten.

Bog arum [1]
Calla palustris (Araceae)

Marginal perennial. Height Up to 30cm. **Habitat** Swampy areas and wet woodlands near ponds. **Distribution** All northern Europe, but introduced in British Isles. **Flowers** June–July. **In gardens** Rich wet soil in water up to 25cm deep, in full sun. Pond margins and bog gardens. **Image size** Life-size.

The creeping bog arum was introduced to gardens in the 19th century, from where it has naturalised in the wild. The mass of greenish-yellow flowers are densely packed into a specialised type of spike called a spadix, which arises from a creamy-white leaf-like spathe. Unlike in the related true *Arum* species (p.209), this does not envelope the flower spike. The flowers are bisexual, except at the top of the spike where they are male only. If fertilised, the spike becomes a striking mass of red berries in autumn. The glossy, dark green leaves are rounded, with distinctly heart-shaped bases. To restrain the creeping rhizomes, plant the bog arum in aquatic planting baskets.

Frogbit [2]
Hydrocharis morsus-ranae (Hydrocharitaceae)

Floating aquatic. Height Stolons 50–100cm long. **Habitat** Ponds, canals and ditches. **Distribution** All northern Europe. **Flowers** July–August. **In gardens** Wildlife ponds. **Image size** Two-thirds life-size.

The simple beauty of this aquatic plant is reflected in its scientific name, from the Greek *hydro* (water) and *charis* (charm). The species name *morsus-ranae* is a translation of the common name frogbit. The pure white flowers, 18–20mm across, each have three petals with a yellow spot at their base and crinkled edges. They float on the water's surface together with leathery kidney-shaped leaves, whose undersides are flushed red. Long roots anchor the buoyant stems.

Frogbit is a popular aquatic for garden wildlife ponds. To guarantee survival, make sure the pool does not freeze solid.

Marsh helleborine [3]
Epipactis palustris (Orchidaceae)

Perennial. Height 15–50cm. **Habitat** Marshes, fens and other damp habitats. **Distribution** Throughout northern Europe except the extreme north, Faeroe Is and Iceland. **Flowers** July–August. **Image size** One and a half life-size.

This beautiful helleborine is often found under cover among tall reeds and hemp agrimony (p.143). The short stems are sturdy, and the alternate sheath-like leaves deeply veined. The flowers have brownish or purplish sepals, a large white lip and crimson-marked upper petals, designed to ensure pollination by visiting bees. A bee landing on the lower lip forces the flower open. As it crawls towards the nectar, the flower closes, trapping the bee, which has to force its way out, rubbing against sticky masses of pollen. When it visits another helleborine, the pollen is transferred to the stigma of its flowers.

Giant hogweed [4]
Heracleum mantegazzianum (Apiaceae or Umbelliferae)

Biennial. Height Up to 5.5m. **Habitat** River-banks, footpaths and agricultural land, especially where damp. **Distribution** Introduction from south-western Asia; now naturalised in northern Europe. **Flowers** June–August. **In gardens** Usually regarded as a pernicious weed. **Image size** One-twenty-fourth life-size.

First cherished in 19th-century gardens for its statuesque habit, giant hogweed has been cast as a villain since it escaped into the countryside. The impressive red-spotted stems can reach up to 10cm in diameter, and are covered in hair-like spines. Enormous umbels – up to 50cm wide – of white 8–20mm flowers produce up to 5,000 seeds. It grows fast, and is vulnerable to normal weedkillers only when young; the roots are difficult to kill. Its sap and bristles contain irritants that blister skin.

It is illegal to cultivate giant hogweed in Britain where it can have an impact on the wild environment. Yet it is better loved in Norway and is known as the Tromso palm.

Wild angelica [5]
Angelica sylvestris (Apiaceae or Umbelliferae)

Perennial. Height 1–2m. **Habitat** Damp, grassy places. **Distribution** Throughout northern Europe. **Flowers** July– October. **In gardens** Moist, fertile soil, full or partial shade. Woodland and wild-flower gardens. **Image size** One-quarter life-size.

Wild angelica is closely related to the garden species (*A. archangelica*) and gives off the same sweet scent. But it is impossible to mistake the two. Wild angelica has majestic purple-stained (instead of all-green) stems; they are both deeply grooved and almost completely hairless. The umbels of pale pink-flushed white flowers can reach 15cm across; garden angelica's are greenish-yellow. The leaves are much divided and have large sheaths that clasp the stem; on the upper part, these sheaths serve to protect the developing flower heads. Wild angelica's tall, architectural stems make a striking impact in a wild-flower garden.

Fool's parsley [6]
Aethusa cynapium (Apiaceae or Umbelliferae)

Biennial. Height Up to 1m. **Habitat** Farm and waste land, open woodland and roadsides. **Distribution** All northern Europe except far north. **Flowers** June–October. **In gardens** An occasional weed. **Image size** One-third life-size.

You would be a fool indeed to eat the deadly poisonous fool's parsley in the mistaken belief that it is true parsley. It contains two alkaloids, aethistine and aethusanol, and if enough is consumed it can be fatal. However, most people are put off by its nauseating smell when crushed. The herbalist John Gerard noted this 'naughtie smell' and the plant's similarity to hemlock (p.59), which is also lethally poisonous. The foliage has a deceptively parsley-like appearance, consisting of lobed, glossy green leaflets. The hollow stems are finely ridged and branched, with umbels of white flowers at their tips – each umbel has 10 to 20 rays. The flowers are quite distinctive, with uneven petals and long bracts, which hang down like threads.

Shaggy soldier [1]

Galinsoga quadriradiata (Asteraceae or Compositae)

Annual. Height 10–70cm. **Habitat** Disturbed ground, arable land and gardens. **Distribution** Britain (S. England, S. Wales). **Flowers** May–October. **In gardens** A weed. **Image size** Two-thirds life-size.

Introduced, with its close relative gallant soldier, (*Galinsoga parviflora*) as a garden plant from South America, shaggy soldier escaped and was soon relegated to weed status. The small (3–5mm) flowers grow on branched stems and mostly consist of a central yellow disc with a thin smattering of white, three-toothed rays. The oval, toothed leaves appear in opposite pairs on hairy stems (the origin of the epithet 'shaggy'). Both species are named after Mariano Martinez Galinsoga, an 18th-century Spanish botanist – whose name was corrupted in Britain to 'gallant soldier'.

Fool's watercress [2]

Apium nodiflorum (Apiaceae or Umbelliferae)

Perennial. Height 30–90cm. **Habitat** Chalky and nutrient-rich soils in wet habitats – still water and moderate streams. **Distribution** Belgium, Britain, France, Holland; naturalised in some other areas. **Flowers** July–August. **Image size** Two-thirds life-size.

Fool's watercress is easily mistaken for both watercress (p.51) and lesser water parsnip (p.59). Its finely grooved, prostrate stems carry loose umbels of white, starry flowers, each of which has, opposite it, a leaf stem of oval, finely toothed leaflets. Water parsnip has more leaflets, with irregular teeth. Fool's and true watercress often grow side-by-side, but the latter's leaflets are wider, with no teeth. In the west of Britain, where people do know the difference, fool's watercress is deliberately picked and cooked with meat dishes and pasties, even though it tastes rather bland.

Meadowsweet [3]

Filipendula ulmaria (Rosaceae)

Perennial. Height 60–120cm. **Habitat** Damp places. **Distribution** Throughout northern Europe. **Flowers** June–September. **In gardens** Damp, fertile soil, full sun, partial shade. Wild-flower and woodland gardens. **Image size** Two-thirds life-size.

The name meadowsweet is a corruption of 'meadwort', referring to the herb's use in flavouring beer and mead rather than its meadow habitat. Cream, almond-scented, 4–8mm, five-petalled flowers appear in dense, fluffy clusters on reddish upright stems. The dark green leaves are deeply toothed and have pale undersides. The whole plant was used to combat stomach complaints and pneumonia. In 1838, salicylic acid was first extracted from both meadowsweet and willow bark. In 1899 the extract was synthesised to make aspirin, a name taken from meadowsweet's old botanical name, *Spirea ulmaria*

There are several clump-forming garden cultivars: 'Aurea' has bright yellow foliage, turning lime-green in summer; 'Variegata' is a striking green striped with yellow.

Wild celery [4]

Apium graveolens (Apiaceae or Umbelliferae)

Biennial. Height 30–100cm. **Habitat** Often brackish soil in damp habitats and on coastlines. **Distribution** Britain and Ireland to Denmark; naturalised in some other parts of southern and central Scandinavia. **Flowers** June–August. **Image size** Two-thirds life-size.

This plant is the ancestor of our cultivated salad celery, and all parts of it have a strong celery smell. In comparison to some of its cousins in the same family, the umbels of white flowers and divided leaflets are quite dainty – unlike the stems, which are chunky, grooved and hollow. The whole plant is used in Indian Ayurvedic medicine for hiccups, asthma and flatulence. It is said to be both a sedative and aphrodisiac, which may or may not have been any use

to the dead Egyptian King Tutenkhamun, in whose tomb it was buried.

The scientific name *Apium* either comes from the Latin for celery and parsnip, or is a Latin variant of the Celtic word for water, *apon*. Both origins are possible, as wild celery thrives in damp places.

Sneezewort [5]

Achillea ptarmica (Asteraceae or Compositae)

Perennial. Height 30–90cm. **Habitat** Damp grassy places. **Distribution** All northern Europe except the northernmost tip. Naturalised in Iceland. **Flowers** July–September. **In gardens** Moist, well-drained soil, full sun. Herbaceous borders, wild-flower gardens. **Image size** Two-thirds life-size.

The 16th-century English herbalist John Gerard recorded that the strong smell of sneezewort flowers cause sneezing. Later a powder made from the dried leaves was recommended as a snuff to clear the head, and chewing the roots was meant to cure toothache. The 12–18mm daisy-like flowers appear at the tips of tall, branched stems and have a greenish-white disc surrounded by white rays. The long, narrow leaves have finely serrated edges and a hot, acrid taste. There are several attractive, double-flowered garden cultivars, among them 'Boule de Neige' and 'The Pearl'.

Autumn lady's tresses [1]

Spiranthes spiralis (Orchidaceae)

Perennial. Height 20cm. **Habitat** Dry grassy habitats. **Distribution** Wales, England and north to Denmark; rare in Ireland. **Flowers** August–September. **Image size** One and a half life-size.

This charming member of the orchid family produces delicate, vanilla-scented white flowers, 6–7mm long, in a spiral like a miniature hair braid. These flower spikes set it and its close relatives apart from other orchids, but this species is even more unusual because they seem to emerge at the side of a rosette of ribbed, oval leaves, rather than from its centre. In fact, the leaves at the base of the spike die before the flowers bloom, but all the while a new rosette is unfurling beside it, ready for next year. Like many orchid flowers, those of autumn lady's tresses are adapted to accommodate pollinating bees.

Round-leaved wintergreen [2]

Pyrola rotundifolia (Pyrolaceae)

Creeping perennial. Height Up to 20cm. **Habitat** Short turf, woodland, bogs, fens and dunes. **Distribution** Patchily throughout northern Europe except Faeroe Is and Iceland. **Flowers** June–September. **In gardens** Well-drained soil in dappled or partial shade. Woodland and rock gardens. **Image size** Half life-size.

The wintergreens are a group of relatively shy plants that have been much admired for their medicinal attributes. They contain an oil, which is now reproduced synthetically to relieve muscular complaints and as a flavouring for confectionery. In round-leaved wintergreen, the ramrod-straight stems carry racemes of 9–12mm white flowers on short pendulous stalks. The flowers are hermaphrodite, possessing both male and female parts; the style is S-shaped and particularly prominent in this species. The glossy toothed leaves are more or less round, with the leaf stems conspicuously longer than the leaf blades. Wintergreens

are difficult to cultivate and don't like being transplanted. Round-leaved wintergreen is, however, one of the few species that adapts relatively easily to gardens – so long as it is left undisturbed.

Field pennycress [3]

Thlapsi arvense (Brassicaceae or Cruciferae)

Annual. Height Up to 60cm. **Habitat** Waste and arable land. **Distribution** All northern Europe except far north. **Flowers** May–August. **Image size** Half life-size.

Everything about field pennycress is geared towards producing and distributing as much seed as possible in a season. Tiny white flowers are arranged in racemes on branched, upright stems, which actually lengthen when the plant is in fruit. The toothed leaves are unstalked and clasp the stem. When the flat, oval fruits – which are 10–15mm across – ripen they turn yellow and are propelled through the air on their prominent wings. Field pennycress gives off a strong and unpleasant smell when crushed – a very effective deterrent against browsing animals. The common name refers to the fruits, which were thought to look like coins.

Japanese knotweed [4]

Fallopia japonica (syns. *Polygonum cuspidatum*, *Reynoutria japonica*; Polygonaceae)

Spreading perennial. Height Up to 2m. **Habitat** Wasteland, railway embankments and damp habitats. **Distribution** Naturalised throughout northern Europe. **Flowers** August–October. **Image size** One-quarter life-size.

Once grown for its ornamental value, Japanese knotweed has become a notorious weed that is extremely difficult to eradicate. It is illegal to plant it in Britain today, but it's easy to understand why it was introduced from the Far East in the first place. The Victorians valued it for its majestic forked, cane-like stems, handsome

heart-shaped leaves up to 30cm long, and elegant sprays of white flowers, which emerge at the leaf axils. Unfortunately once it escapes, it forms impenetrable thickets that swamp anything in its path.

Common sundew [5]

Drosera rotundifolia (Droseraceae)

Perennial. Height Up to 25cm. **Habitat** Wet heaths, sphagnum bogs and moors. **Distribution** All northern Europe. **Flowers** June–August. **Image size** Life-size.

This round-leaved sundew grows in wetter conditions than its long-leaved relation, *D. intermedia* – usually on sphagnum moss – and may be seen floating on small ponds. Both are carnivorous plants with a taste for small flies and have leaves covered with the same kind of glandular hairs and attractive silver-headed glands. Common sundew's flowers are white, usually with six petals. In Britain they are almost always self-pollinating.

Long-leaved sundew [6]

Drosera intermedia (Droseraceae)

Perennial. Height Up to 10cm. **Habitat** Bogs, heathland and moors. **Distribution** All northern Europe except Iceland and Faeroe Is. **Flowers** June–August. **Image size** Life-size.

Sundews grow on soils where there is little or no nitrogen and other vital minerals, so they have evolved a mechanism for getting these by trapping flies. The oblong leaves are covered in glandular hairs, each with a shining blob at the end. These blobs look like dewdrops – hence the common name. The blobs attract small flies, which probably perceive them as water drops where they can lay eggs. They stick to the blob, and the leaf rolls over and traps them. In summer, the spikes of white flowers, with up to eight petals, unroll. British sundews are usually self-pollinating so occasionally the flowers will not open.

Dittander [1]

Lepidium latifolium (Brassicaceae or Cruciferae)

Perennial. Height Up to 1.3m. **Habitat** Damp coastal habitats, including salt marshes. **Distribution** All northern Europe except Iceland, Faeroe Is, Norway and Finland. Rare in Ireland. **Flowers** June–July. **Image size** Life-size.

Wild dittander is scarce in many areas, but it was cultivated for centuries for its hot, peppery roots. They were used to make a horseradish-like condiment and in Germany (where it is called *Pfefferkraut*) as a flavouring before pepper was introduced. The plant has small white flowers in dense clusters at the top of the stem. The four sepals have white margins and are half the length of the white petals. The blue-green lower leaves are large and long-stalked, with toothed edges; farther up the stem the leaves have no stalks and are narrower.

Dittander was used for the treatment of leprous sores, and some plants still grow on the site of old hospitals. It is available from a few specialist nurseries, but beware: it is liable to spread vigorously, so is best contained in pots.

Sea mayweed [2]

Tripleurospermum maritimum (syn. *Matricaria maritima*; Asteraceae or Compositae)

Biennial or perennial. Height 60cm. **Habitat** Coastal habitats. **Distribution** All northern Europe. **Flowers** July–September. **Image size** Life-size.

Sea mayweed belongs to a confusing group of plants. Some, such as pineappleweed (p.111), have a distinctive fruity smell when bruised; others, such as this species, have none. Its white daisy-like 15–50mm flower heads have white outer florets around a yellow disc. Unlike those of other members of the family, its seeds have no white parachute of fine hairs to aid wind distribution. The leaves are divided into narrow segments and are slightly succulent; as a result they can retain moisture and tolerate harsh, drying seaside conditions.

Scentless mayweed (*T. inodorum* or *M. perforata*; p.64) is very similar; it inhabits coastal locations but is found inland too. The two species hybridise and produce fertile seeds.

Small teasel [3]

Dipsacus pilosus (Dipsacaceae)

Biennial. Height 1.5m. **Habitat** Damp woods, stream edges and hedge-banks on chalky soil. **Distribution** Northern Europe to Denmark, excluding Ireland. **Flowers** August–September. **Image size** Life-size.

Like the wild teasel (*D. fullonum*, p.202), this is a prickly customer. The slender, upright stems are covered with bristles from top to bottom; their tips end in soft hairs, hence the name *pilosus* (literally, 'covered in soft hairs'). The oval, toothed leaves are prickly too, but only on the midrib below; at the base of the stem they form a rosette. The white pincushion-like flowers, 6–9mm long, have pinkish stamens and look more like a scabious (also a relative; see p.182, 185, 193) than a typical teasel.

Stone parsley [4]

Sison amomum (Apiaceae or Umbelliferae)

Biennial. Height Up to 1m. **Habitat** Roadsides, hedge-banks and grassy or cultivated areas on clay or chalky soil. **Distribution** Northern Europe as far north as France and northern England; not in Ireland. **Flowers** July–September. **Image size** Half life-size.

Stone parsley is easily recognised by the smell of its crushed leaves – once you know. It has been described as smelling like petrol. Others say it is like petrol mixed with nutmeg, but it can really only be described as the smell of stone parsley!

It is an upright, often purple-tinged plant. The lower leaves have long stalks and are divided into paired, lobed leaflets, with fine teeth. The stem leaves usually have three lobes, each with toothed segments.

The flower head forms a narrow umbrella of three to six rays of differing lengths. Each is topped by a small umbel of up to six flowers, with five green-white petals.

Common chickweed [5]

Stellaria media (Caryophyllaceae)

Sprawling annual. Height Stems up to 50cm long. **Habitat** Bare, waste and cultivated land, roadsides and beaches. **Distribution** All northern Europe. **Flowers** All year. **In gardens** A common and persistent weed. **Image size** Life-size.

Common chickweed is a master of survival. Once the seed has germinated, young plants grow at great speed, flower for as long as possible and then set hundreds of new seeds to guarantee the next generation. It is a sprawling plant that most people will recognise as a pretty coloniser of road verges, yet it is very tenacious and, once it starts growing in the garden, it is almost impossible to eradicate. The tiny star-like flowers – the genus name *Stellaria* means 'little star' – have five petals so deeply notched that they appear to be ten. The oval leaves have faintly wavy margins and are arranged in opposite pairs. A single row of hairs appears along one side of the stems. Drops of water collect on these hairs and are then absorbed by the plant – a clever survival technique that gives common chickweed the competitive edge. The common name, chickweed, is a reference to it being a favourite food of fowls.

White helleborine [1]

Cephalanthera damasonium (Orchidaceae)

Perennial. Height Up to 60cm. **Habitat** Beech woodland and other shady habitats. **Distribution** Most of northern Europe as far north as southern Sweden, except Ireland and Faeroe Is. **Flowers** May–July. **Image size** Two-thirds life-size.

White helleborine is, like many orchids, a shy plant that favours the shelter of beech trees where little else grows. The flowers are white or cream, and almost always look as if they are only half-open. The top two of the inner perianth segments ('petals') are similar to the outer ones, and the third forms the lip, which has yellow-orange ridges. The flowers attract pollinating insects by secreting syrup. Their tubular form helps to draw insects into contact with the pollinia but, for all that, most flowers seem to self-pollinate. The foliage is bluish-green; the alternate broad lanceolate to oval leaves have no stalks and clasp the angled stem.

Commom comfrey [2]

Symphytum officinale (Boraginaceae)

Perennial. Height Up to 1.5m. **Habitat** Damp meadows, fens, marshes, river-banks and roadsides. **Distribution** Most of northern Europe but naturalised in Ireland and much of the far north. **Flowers** May–July. **In gardens** Moist, fertile soil in full sun or partial shade. Shady borders, and woodland and vegetable gardens. **Image size** One quarter life-size.

Common comfrey is a handsome plant with sturdy upright stems and attractive oval leaves, all covered in rough bristles. The stem leaves have no stalks, the leaf margins merging with the stem and forming wings. The flowers may be whitish, creamy-yellow, purple or pinkish, and are arranged in a coiled spray called a cyme. Wily gardeners compost the leaves or steep them in water to make a powerful liquid fertiliser with a particularly unpleasant smell. They can be boiled like spinach or fried in batter to make a Bavarian delicacy called *Schwarzwurz*.

Water soldier [3]

Stratiotes aloides (Hydrocharitaceae)

Floating aquatic perennial. Height Up to 50cm above water level. **Habitat** Ponds, dykes and canals. **Distribution** Throughout northern Europe. Introduced in Ireland. **Flowers** June–August. **In gardens** Water 30cm or more deep in full sun. Pools. **Image size** One-third life-size.

Water soldier's handsome sword-like leaves (in rosettes rather like pineapple tops) and exotic white flowers would seem more at home in the tropics than in our canals and ponds. The leaves have sharp, serrated edges, and the three-petalled 3–4cm blooms nestle among them, just above the water surface. Male and female flowers appear on separate plants – the females solitary, the males (very rare in Britain) in clusters. But the plant spreads vigorously by runners, remaining submerged except when flowering. It makes a striking addition to a garden pool, so long as it is kept under control.

Shepherd's purse [4]

Capsella bursa-pastoris (Brassicaceae or Cruciferae)

Annual. Height Up to 50cm. **Habitat** Cultivated, bare ground and wasteland. **Distribution** All northern Europe. **Flowers** All year. **In gardens** A common weed. **Image size** Half life-size.

The 'purses' of the attractive country name are the delicate, 6–9mm heart-shaped seedpods, which are shaped like the leather purses that shepherds used for carrying their daily food. Despite its rather fragile appearance, shepherd's purse is one of the most effective competitors of the wild-flower community. Self-pollinating, it produces hundreds of seeds throughout the year; they live for long periods in the soil and germinate on cultivated or bare ground at any time. The flower spikes arise from rosettes of deeply lobed leaves and carry clusters of tiny white flowers at the tips. To eradicate the plants from your garden, simply pull them out by hand – they come up easily – or use a hoe.

Grass of Parnassus [5]

Parnassia palustris (Parnassiaceae or Saxifragaceae)

Perennial. Height Up to 30cm. **Habitat** Marshes, wet pastureland and fens. **Distribution** All northern Europe except far north. **Flowers** June–September. **In gardens** Wet soil in full sun. Bog gardens. **Image size** Half life-size.

The solitary white flowers grow on stout fleshy stems above a mass of heart-shaped leaves. The Swedish botanist Linnaeus thought it so beautiful that he named it after Mount Parnassus, in Greece. In reality, however, grass of Parnassus is not a grass – it was once considered a relative of the saxifrages but is now placed in its own family – and it doesn't hail from a Greek mountain. The 15–30mm flowers have five petals, each with a tracery of delicate green veins, and five green sepals. The fringed modified stamens at the centre are tipped with yellow nectar-bearing glands, which give off a honey-like scent irresistible to pollinating insects.

Mountain avens [6]

Dryas octopetala (Rosaceae)

Evergreen sub-shrub. Height Up to 50cm. **Habitat** Rock crevices, mountain ledges and cliffs. **Distribution** Britain, Ireland, France, Germany and Scandinavia. **Flowers** May–July. **In gardens** Well-drained, gritty soil in sun or partial shade. Ground cover, rock gardens and walls. **Image size** One-third life-size.

The captivating and apparently so-delicate mountain avens is, in fact, a tough customer that has been around since before the last Ice Age. The cup-like flowers are snow-white, usually with eight rounded petals, with a coronet of shimmering golden stamens at their centre. They arise on long stems from a mat of neat, evergreen foliage with silvery undersides. Each scalloped leaf is like a miniature oak leaf – the genus name *Dryas* is from the Greek *drus* (oak). Mountain avens is relatively elusive, but it flourishes on the limestone pavements of the Burren in Ireland in May.

Old man's beard, traveller's joy [1]
Clematis vitalba (Ranunculaceae)

Woody climber. Height up to 30m. **Distribution** Hedgerows, woodland and scrub in Britain, France, Germany, Holland and Belgium. Introduced to Ireland and naturalised in southern Scandinavia. **Flowering period** July-September. **Image size** Half life-size.

The twining, flaky stems of old man's beard, which habitually reach Amazonian proportions, drape themselves jungle-like over trees and hedges. The 16th-century herbalist John Gerard christened it traveller's joy because it 'deck[s] and adorn[s] waies and hedges, where people trauell'. The name old man's beard comes from the hairy seed heads (see smaller photograph), which last from autumn well into winter and are designed to float on the wind. The fragrant white summer flowers, 18–20mm across, give it yet another poetic name, virgin's bower. They consist of four hairy sepals and a froth of stamens. The leaves comprise three or five oval, toothed leaflets on supple, winding stems.

The idea that the woody stems could be smoked like cigars spread across Europe. As a result, in Britain it was known as gipsies' (or boys') bacca, in Germany, *Rauchholz*, France, *bois à fumer* and Holland, *smookhout*.

Dewberry [2]
Rubus caesius (Rosaceae)

Sub-shrub. Height Up to 30cm. **Habitat** Damp grassland, scrub, hedge-banks and dunes. **Distribution** All northern Europe except far north. **Flowers** May–September. **Image size** Life-size.

Dewberry's sprawling stems are similar to those of brambles (p.139), but are weaker, round and less prickly. Each leaf consists of three oval, toothed leaflets instead of five. The 20–25mm flowers are white, or occasionally pink, with five rounded petals and narrow sepals, which are longer than the petals. The deep purple fruits have a

blue-grey bloom, like that on sloes and damsons. They have fewer, larger individual segments than blackberries, but are just as juicy and delicious to eat.

Field rose [3]
Rosa arvensis (Rosaceae)

Scrambling shrub. Height Up to 3m. **Habitat** Woodland margins, hedgerows and scrubland. **Distribution** Northern Europe as far north as England and Wales, Ireland, Holland and parts of Germany. **Flowers** June–August. **Image size** Life-size.

The name field rose is something of a misnomer – it tends to grow in woodland rather than open fields. Similar to the dog rose (p.136), the field rose always produces pure white flowers with startling yellow stamens and the warm scent of musk. Like those of dog roses, the arching stems are armed with red hooked thorns, which support the field rose as it scrambles among hedgerows and shrubs. The flowers are followed by attractive scarlet, oval hips in autumn. Although its flowers are too fleeting to make it of much garden value, it was a parent of the Ayrshire rose cultivars such as 'Bennett's Seedling' and 'Dundee Rambler', which were popular a century ago.

Stone or rock bramble [4]
Rubus saxatilis (Rosaceae)

Creeping perennial. Height Up to 40cm. **Habitat** Woodland, scrub and shady, rocky habitats, mostly in hills. **Distribution** All northern Europe. **Flowers** June–August. **Image size** One and a half life-size.

A diminutive close relative of the bramble or blackberry (p.139), stone bramble spreads on creeping stems covered their length in prickles. Each leaf is made up of three oval leaflets with toothed edges. The flowering stems are upright and carry only a few pure white flowers with slender, upright petals and reflexed sepals. They are followed by edible scarlet fruits in the autumn. Unlike blackberries and raspberries (*Rubus idaeus*), the fruits of stone brambles have only a few juicy segments.

Wild tulip [1]

Tulipa sylvestris (Liliaceae)

Bulbous perennial. Height Up to 50cm. **Habitat** Dry grassy places and copses. **Distribution** Native to parts of France; naturalised elsewhere as far north as southern Sweden. **Flowers** April–May. **In gardens** Any well-drained soil, preferably with hot sun in summer. Beds and borders. **Image size** One-third life-size.

Outside its natural range, the wild tulip often indicates the location of a very old garden that has reverted to nature. It has been cultivated since the 16th century and was first recorded in Britain as naturalised in the late 17th. It is a handsome plant with blue-grey narrow leaves and yellow, scented flowers – one or two to a stem – that droop when in bud. Each has six pointed segments, the outer ones often tinged green. They open wide in full sun, and are pollinated by small insects, but rarely seem to set seed.

Winter aconite [2]

Eranthis hyemalis (Ranunculaceae)

Tuberous perennial. Height Up to 15cm. **Habitat** Damp woodland and banks. **Distribution** Widely naturalised in northern Europe except Ireland and most of Scandinavia. **Flowers** January–March. **In gardens** Any well-drained soil in partial or dappled shade. Good around trees. **Image size** One-and-a-half life-size.

The cheerful, bright yellow 2–3cm flowers of winter aconite appear on bare stems before the main root leaves – which grow only as the flowers are setting seed. The flowers have six petal-like sepals, with a ruff of leaf-like bracts beneath, and open only when temperatures are high enough to keep pollinating insects active. A southern European native, winter aconite was introduced as a garden plant before 1600 and escaped into the wild. The tubers may be difficult to establish in the garden; if so, try planting growing plants in spring. Once naturalised, it can form large colonies. *E. cilicica* is larger-flowered, good in gardens.

Wild daffodil, Lent lily [3]

Narcissus pseudonarcissus (Amaryllidaceae or Liliaceae)

Bulbous perennial. Height 15-35cm. **Habitat** Deciduous woodland and meadows. **Distribution** Britain, France, Germany, Holland, naturalised elsewhere. **Flowers** February –March. **In gardens** Well-drained, moist soil. Woodland gardens and grassy areas. **Image size** Two-thirds life-size.

The nodding golden trumpets of wild daffodils are sadly now a rarity in their native habitats, mostly because of woodland clearance, intensive farming and the indiscriminate uprooting of the bulbs in the past. Today the species is protected in Britain, and grows only sporadically from Sussex to the Lake District. Tightly packed buds unfurl to reveal opaque papery tepals, which vary from pale yellow to cream and envelop the deep yellow trumpet or corona. The double-angled, blue-green stems are fleshy and upright, with masses of strap-like leaves emerging from their base.

Lesser celandine [4]

Ranunculus ficaria (Ranunculaceae)

Perennial. Height Up to 25cm. **Habitat** Damp grassy habitats, woodland, hedgerows, and stream- and river-banks. **Distribution** All northern Europe except far north. **Flowers** February–May. **In gardens** Humus-rich moist soil in semi- or full shade. Shady borders and woodland gardens. **Image size** Two-thirds life-size.

Only dogged curmudgeons will fail to respond to a carpet of lesser celandine. After the long winter months, the starry flowers, each with up to 12 petals and three sepals, illuminate the undergrowth with their glossy golden sheen – except when they close up in cloudy weather or at dusk. One flower grows on each stem that rises from a chaotic clump of heart- or ivy-shaped leaves. The flowers are pollinated by insects, but the plant also multiplies by tiny bulbils that nestle in the leaf axils and fall to the ground in early summer. Lesser celandine spreads vigorously; keep it under control by pulling out unwanted patches by hand.

Opposite-leaved golden saxifrage [5]

Chrysosplenium oppositifolium (Saxifragaceae)

Creeping perennial. Height 5–15cm. **Habitat** Wet or damp habitats, mostly on acid soil. **Distribution** All northern Europe except Iceland and Faeroe Is. **Flowers** March–July. **In gardens** Moist, humus-rich soil in shade. Ground cover in bog gardens, woodland gardens and shady borders. **Image size** Two-thirds life-size.

This diminutive plant paves the ground with golden speckles as it spreads, despite its tiny flowers having no petals. Instead, they each have eight bright yellow stamens, which are surrounded by four yellowish-green sepals. The round leaves are edged with blunt teeth and are arranged in opposite pairs – distinguishing it from the very similar alternate-leaved golden saxifrage (*Chrysosplenium alternifolium*).

Coltsfoot [6]

Tussilago farfara (Asteraceae or Compositae)

Creeping perennial. Height 15cm. **Habitat** Damp habitats, roadsides, arable land, woodland margins and shingle. **Distribution** All northern Europe. **Flowers** February–April. **Image size** Two-thirds life-size.

Each stout, scaly coltsfoot stem carries a single 30mm flower head whose yellow disc florets are surrounded by a halo of narrow, frilly ray florets. They appear well before the leaves, standing to attention like miniature golden chimney-brushes, to enliven the dull, cold days of late winter. The heart- or hoof-shaped leaves, up to 30cm across, give the plant its British common name; they emerge at the base only after the flowers have died. The leaf was dried and smoked as herbal tobacco to relieve asthma and coughs. The botanical name *Tussilago*, meaning 'cough dispeller', dates back to Roman times. Coltsfoot spreads far and wide. Whether you regard it as a weed or tolerate it for its late winter colour, keep it under control by removing the flower heads before seeds are set. If it gets too big for its boots, pull the roots out by hand.

Yellow figwort [1]
Scrophularia vernalis (Scrophulariaceae)

Perennial. Height Up to 50cm. **Habitat** Shady places, wasteland and mountainous woods. **Distribution** France and southern Germany. Introduced to Britain, Belgium, Denmark, southern Sweden and Holland. **Flowers** April–June. **Image size** Two-thirds life-size.

In most ways, the yellow figwort is typical of other *Scrophularia* species, with its hairy square stems and sharply toothed heart-shaped leaves arranged in opposite pairs. One difference is the greenish-yellow flowers: instead of two lips, they have five more or less equal lobes with pointed rather than rounded sepals. Despite having no scent, they attract pollinating bees.

Globe flower [2]
Trollius europaeus (Ranunculaceae)

Perennial. Height 40–70cm. **Habitat** Damp grassland, woodland and stream-banks. **Distribution** Most of northern Europe; in Britain mainly the north and west. **Flowers** May–August. **In gardens** Moist, fertile soil in full sun or partial shade. Bog gardens, damp borders and pool margins. **Image size** Two-thirds life-size.

This plant's glorious globes of yellow flowers – each up to 12cm across – are in fact constructed from layers of up to 15 shiny, petal-like sepals. The true petals are reduced to a whorl of nectaries hidden in the centre. The deeply cut leaves, which are more or less stalkless, are typical of the buttercup family. Like several of its relations, the globe flower is poisonous, but there are some striking cultivars for the garden, including the lemon-yellow 'Canary Bird'.

Rape [3]
Brassica napus (Brassicaceae or Cruciferae)

Annual or biennial. Height Up to 1.5m. **Habitat** Arable land, roadsides and wasteland. **Distribution** Introduced in most of northern Europe. **Flowers** May–August. **Image size** Two-thirds life-size.

Vast luminous fields of rape are now a familiar early summer spectacle. The form usually grown for its oil-bearing seeds is the slender-rooted oilseed rape, *B. napus* subsp. *oleifera*; the wild-type *B. napus* has a carrot-like root. The flowers, with four pale yellow petals, appear in clusters at the tip of bluish-grey stems; they are followed by long, thin, beaked pods, which contain the oily seeds. The leathery leaves, with a network of craggy veins, have no stalks and clasp the stem.

Alexanders [4]
Smyrnium olusatrum (Apiaceae or Umbelliferae)

Biennial. Height Up to 1.5m. **Habitat** Roadsides, wasteland, cliffs and woodland margins, especially near coasts. **Distribution** South from north-western France. Introduced to Britain, Ireland and Holland. **Flowers** April–June. **Image size** Two-thirds life-size.

Alexanders smells and tastes like wild celery (p.71), but its umbels of tiny greenish-yellow flowers on stiff, statuesque stems are reminiscent of wild angelica (p.68). Each leaf consists of three oval, toothed leaflets, whereas angelica has many finely divided leaflets in rows.

The name refers to Alexander the Great, from whose homeland of Macedonia the

plant originally hailed. The Romans probably introduced it as a potherb, and every part has been eaten over the centuries. Today the young leaves are eaten in salads and the young stems treated like asparagus.

Wild cabbage [5]
Brassica oleracea (Brassicaceae or Cruciferae)

Biennial or perennial. Height Up to 2m. **Habitat** Coastal habitats, cliffs and roadsides. **Distribution** Britain and France. Naturalised in Germany and Holland. **Flowers** May–September. **Image size** Two-thirds life-size.

The voluminous pale yellow flowers of the wild cabbage contrast beautifully with its shiny, plum-coloured stems, which become woody with age. The leathery leaves have deeply textured surfaces and undulating edges, and are stalkless towards the top of the stem. Cabbage has been cultivated since Greek and Roman times, and is thought to be the ancestor of our garden broccoli, Brussels sprouts and cauliflower, as well as of modern cabbage varieties. Unlike these, however, the leaves of wild cabbage are very bitter unless you boil them at least twice.

Bermuda buttercup [6]
Oxalis pes-caprae (Oxalidaceae)

Perennial. Height 30cm. **Habitat** Waste and arable land. **Distribution** Introduced to Britain and France. **Flowers** March–June. **Image size** Two-thirds life-size.

The Bermuda buttercup is not a buttercup, nor is it from Bermuda. It is, in fact, a South African native, first introduced as an exotic glasshouse plant. Now it is happily settled in the warmer parts of Britain and France – in the Isles of Scilly, especially. The Bermuda buttercup is the largest of all the *Oxalis* species found here. The lemon-yellow flowers, arranged in clusters on spindly stems, have five petals and reach up to 25mm across once they elegantly unfurl from pointed spiral buds.

Common dandelion [1]

Taraxacum officinale (Asteraceae or Compositae)

Perennial. Height 5–40cm. **Habitat** Cultivated land and grassland. **Distribution** All northern Europe. **Flowers** March–October. **In gardens** A well-known weed. **Image size** One-sixth life-size.

Over 1,000 dandelion species grow in Europe, many indistinguishable to an untrained eye. But everyone knows the common dandelion. Its bright yellow flowers consist of layers of strap-shaped florets and open wide in the sun but fold up at night or in dull weather. Each stands on a hollow, fleshy stem, which brims with sticky white sap. Then come conspicuous, fluffy spherical seed heads – 'dandelion clocks' to children. The long leaves are arranged in a basal rosette and have deep triangular lobes that give the plant its common name, from the French *dents de lion* (lion's teeth). Dandelions were used medicinally for constipation and eczema and have plenty of culinary uses too, from wine made from the flower heads to tangy leaf salads.

Leopard's bane [2]

Doronicum pardalianches (Asteraceae or Compositae)

Perennial. Height Up to 80cm. **Habitat** Woods, hedgerows and waste and cultivated land, especially on chalky soils. **Distribution** Belgium, France and Germany. Introduced in Britain. **Flowers** May–July. **In gardens** Easy to grow in most soils, in sun or shade. Herbaceous beds and borders. **Image size** Two-thirds life-size.

This upright plant produces clusters of yellow 30–50mm flower heads above low clumps of hairy heart-shaped leaves. The basal leaves grow in a rosette and have long stalks; up the stem the stalks are shorter until the topmost leaves are stalkless and clasp the stem. The disc florets produce seeds with a parachute of hairs, while the ray florets' seeds have none. The plant is poisonous, and the English name comes from its use, mixed with a chunk of flesh, for poisoning leopards and other wild animals.

Wallflower [3]

Erysimum cheiri (syn. *Cheiranthus cheiri*; Brassicaceae or Cruciferae)

Perennial. Height Up to 50cm. **Habitat** Old walls, cliffs and other dry areas. **Distribution** Widely naturalised in Britain, Ireland, Belgium, Holland, France and Germany. **Flowers** April–June. **In gardens** Any soil in sun. Beds and borders. **Image size** Half life-size.

Originally from the eastern Mediterranean, wallflowers are classic spring bedding plants, whether in cottage gardens or municipal parks. The wild wallflower can best be seen perched on old walls, or in other dry areas. It is a well-branched plant, with four-petalled flowers slightly smaller than in the cultivated forms; they are usually yellow, but often red, purple or brown. The long cylindrical seedpods are hairy and open from the base upwards. Wallflowers have been grown in gardens (often as biennials) since mediaeval times.

Marsh marigold, kingcup [4]

Caltha palustris (Ranunculaceae)

Perennial. Height 10–40cm. **Habitat** Wet habitats. **Distribution** All northern Europe. **Flowers** March–August. **In gardens** Rich, boggy soil in full sun. Pond edges and bog gardens. **Image size** Life-size.

This variable plant is as happy at the edge of water as it is up to its ankles in it. The plump, fleshy stems carry fat round buds that unfurl into dazzling goblet flowers formed by shiny yellow sepals rather than petals. Up to 100 stamens may be crammed into the centre of each flower. The heart-shaped leaves at least double their size once the plant has flowered.

The whole plant is poisonous, but the buds were once pickled and used like capers. In Ireland bunches of marsh marigold are hung over doors on May Day to protect the fertility of cattle from marauding witches and fairies. Among worthwhile varieties for gardens is the double-flowered 'Flore Pleno'.

Primrose [5]

Primula vulgaris (Primulaceae)

Perennial. Height 5–15cm. **Habitat** Shady habitats, hedgerows and grassy banks. **Distribution** N. Europe except Finland and Iceland. **Flowers** February–May. **In gardens** Moist, humus-rich soil in partial shade. Borders, woodland gardens and containers. **Image size** Life-size.

Primrose literally means 'first rose' and refers to this plant's role as the harbinger of spring. The familiar pale yellow flowers emerge from a rosette of soft, wrinkled, oval and slightly toothed leaves. A solitary flower, with notched petals, appears at the tip of each flimsy pinkish stalk. Like many *Primula* species, there are two types of flower on each plant, with the stigma above the anthers ('pin-eyed'), or vice-versa ('thrum-eyed'); this helps to ensure cross-fertilisation. Primrose had many medicinal uses and it is still found in expectorants for bronchitis. Add primrose flowers to green salad leaves for a delightful spring salad.

Cowslip [6]

Primula veris (Primulaceae)

Perennial. Height 10–30cm. **Habitat** Grassland, open woodland and roadsides. **Distribution** All northern Europe except extreme north and Iceland. **Flowers** April–May. **In gardens** Moist, well-drained soil in full sun or partial shade. Borders, lawns and wild-flower meadows. **Image size** Life-size.

The once-common glorious drifts of cowslips became a rare sight in the mid-20th century, mostly because intensive farming did away with their grassland habitats. Now they have made a vigorous comeback along motorway embankments and roadsides. The flowers are held in nodding clusters on hairy stems. Each has five petals that join together to form a tube. The stems emerge from a rosette of deeply veined, wrinkly leaves. Like the primrose (see above), with which it sometimes hybridises, the cowslip has two flower types – 'thrum-eyed' and 'pin-eyed'. The flowers make wonderful wine.

Hoary cinquefoil [1]

Potentilla argentea (Rosaceae)

Perennial. Height Up to 30cm. **Habitat** Dry, sandy grassland, roadsides and coastal habitats. **Distribution** All northern Europe except extreme north, Iceland, Ireland and northern Scotland. **Flowers** June–September. **In gardens** Well-drained soil, full sun. Rock gardens. **Image size** Life-size.

The pretty hoary cinquefoil belongs to a genus of about 500 species that include perennials and shrubs. The names 'hoary' and *argentea* (silver) refer to the whitish felty undersides of the leaves, which are deeply toothed and five-fingered. The leaves are the main way of distinguishing this species from other cinquefoils with similarly divided leaves. The pale yellow flowers are carried on downy, branched stems; they have five petals with slightly scalloped edges.

Yellow pimpernel [2]

Lysimachia nemorum (Primulaceae)

Creeping perennial. Height Up to 40cm. **Habitat** Damp shady habitats: deciduous woodland and shady streamsides. **Distribution** All northern Europe except Faeroe Is and Iceland. **Flowers** June–August. **Image size** Life-size.

Sprawling along the woodland floor on slender stems, the yellow pimpernel turns its solitary yellow flowers towards the shafts of sunlight that pierce the tree canopy. The 12mm flowers are held on long stalks which emerge at the leaf nodes; each flower has five petals with long, pointed sepals behind. The shiny, oval leaves are arranged in pairs and are deeply veined. The yellow pimpernel is closely related to creeping Jenny (p.107) – a bigger plant, popular for ground cover in gardens, that has more rounded leaves with glands on their undersides, and larger flowers with broader sepals. Four centuries ago, both plants were used in medicine to relieve pain.

Roseroot [3]

Sedum rosea (Crassulaceae)

Perennial. Height 20–35cm. **Habitat** Cliffs and rocky places. **Distribution** Northern Britain, Wales, Ireland and Scandinavia; rare farther south. **Flowers** May–June. **In gardens** Well-drained fertile soil in full sun. Rock gardens. **Image size** One-quarter life-size.

It is the surprising damask-rose scent of the cut roots that gives this plant its common name. The fleshy stems carry grey-green, paddle-shaped leaves, which are stiff and succulent, and tinged with purple. The star-like, dull yellow flowers have prominent purple anthers, and sometimes no petals. They are arranged in dense clusters, which nestle among the toothed leaves at the stem tips. Male and female flowers are on separate plants. In Wales roseroot is known as the Snowdon rose (large populations grow on the mountain) and in Ireland 'hero's plant'. Cottage gardeners grew it to make rose-scented water.

Procumbent yellow sorrel, yellow oxalis [4]

Oxalis corniculata (Oxalidaceae)

Creeping perennial. Height Up to 50cm. **Habitat** Dry habitats; wasteland. **Distribution** Introduced to most of northern Europe except Faeroe Is and Iceland. **Flowers** May–October. **In gardens** An invasive, prolific weed. **Image size** Two-thirds life-size.

The procumbent yellow sorrel is a non-native plant with a reputation as a pernicious weed. Yet it's impossible not to be charmed by the pretty, clear yellow flowers with their five rounded petals. They mature into long, hairy seed capsules, which burst open releasing their seed with gusto. The clover-like leaves, with three leaflets, are light-sensitive and fold back at night or in dim light by day. Although poisonous in large quantities, the leaves can liven up an otherwise ordinary salad.

It is the plant's creeping stems and deep taproots that make it so tenacious as a weed.

Slender trefoil [5]

Trifolium micranthum (Fabaceae or Leguminosae)

Annual. Height Stems up to 25cm long. **Habitat** Short grassland, especially near coasts. **Distribution** Western Europe, north to Britain and Norway. **Flowers** June–July. **In gardens** Often a lawn weed. **Image size** Twice life-size.

This small yellow clover is distinguished from its relatives by the upright petal, or standard, of each flower being deeply notched. The pedicels (flower stalks) are slim and about the same length as the calyx tubes. Slender trefoil grows on neutral or slightly acid soils, in places rich in wild flowers, or in drought-prone meadows. It is particularly at home in a lawn, where it provides a feeding ground for small insects. You can keep it under control by raking it out every now and again.

Yellow rattle [6]

Rhinanthus minor (Scrophulariaceae)

Annual. Height 20–50cm. **Habitat** Open grassland, fens, heaths and roadsides. **Distribution** All northern Europe. **Flowers** May–September. **Image size** Life-size.

Like other *Rhinanthus* species, this plant is a hemi- (or semi-) parasite. This means that it helps itself to minerals and water absorbed by other plants, using root-like organs called haustoria – in this case usually from nearby grasses. The clear yellow flowers are tubular, with two lips that part to reveal a pair of violet teeth on the upper lip. The name *Rhinanthus* comes from Greek words meaning 'nose flower'; if you turn the flower upside down, the upper lip looks like a nose. The flower base is encased in a distinctive, inflated pale green calyx, which later protects the ripening fruit and forms the rounded seed capsule. The rattling of ripe seeds inside the capsule is one of nature's alarm clocks, and was used to signal the start of haymaking. Opposite pairs of dark green leaves are arranged all the way up the black-spotted upright stems. They have no stalks and are sharply toothed.

Mouse-ear hawkweed [1]

Pilosella officinarum (syn. *Hieracium pilosella*; Asteraceae or Compositae)

Perennial. Height Up to 30cm. **Habitat** Grassland, wasteland, sand dunes and walls. **Distribution** All northern Europe except far north. **Flowers** May–September. **Image size** Two-thirds life-size.

At a glance, mouse-ear hawkweed looks rather like a common dandelion (p.84), but the solitary lemon-yellow flower heads grow on much slimmer leafless stems, which rise from slender runners, while the basal leaves have no lobes. The stiff blue-grey leaves grow in a loose rosette and are smothered in woolly white hairs – hence the name mouse-ear. Occasionally the outer ray florets are striped beneath with red. *Pilosellas* were once classified with true hawkweeds (*Hieracium* species; p.99), but can be easily distinguished by their leafless stems and creeping stolons.

Stinking iris [2]

Iris foetidissima (Iridaceae)

Rhizomatous perennial. Height Up to 70cm. **Habitat** Hedgerows, woods and sea cliffs, usually on chalky soil. **Distribution** Europe as far north as Britain and France. **Flowers** June–July. **In gardens** Any good soil, preferably in shade. Beds and borders; woodland gardens. **Image size** Half life-size.

No attractive plant deserves an ugly name, but this iris has plenty of others. They include gladdon, gladwyn (an old word for a sword, referring to the leaves) and adder's meat (a reference to the markings on the petals). It is also called the roast beef plant, referring to the smell of the crushed leaves – which is also responsible for the 'stinking' epithet. The flowers are usually bluish-purple, flushed with yellow, but may be a yellowish-beige; the petals have delicate veins. They usually last only a day and it is easy to miss them flowering. The large green fruits open out to reveal orange-red seeds, which remain well into the winter.

Woad [3]

Isatis tinctoria (Brassicaceae or Cruciferae)

Biennial. Height 0.8–1.5m. **Habitat** Dry habitats, rocks and wasteland. **Distribution** Introduced to northern Europe except Ireland, Faeroe Is and Iceland. Rare in Britain and Ireland. **Flowers** July–August. **In gardens** Moist, well-drained, fertile soil in full sun. Wild-flower and herb gardens. **Image size** Two-thirds life-size.

Although not a native, woad has made an indelible impression since it arrived from eastern Europe and western Asia over 1,000 years ago. Its original value was in the blue dye made from it – albeit through a complicated process of pulping, drying, wetting and fermenting. It was eventually superseded by indigo – which in turn was replaced by synthetic dyes.

Woad's fluffy heads of sulphur-yellow flowers are carried on blue-grey branched stems along with substantial foliage, which is distinctly arrow-shaped and shiny. The flowers are followed by pendulous fruits that turn dark brown as they dry out. Although woad is no longer used to make dye, it is admired for its ornamental attributes, which are best appreciated when it is planted en masse. It is particularly attractive to bees, so is perfect for wild-flower gardens.

Hairy buttercup [4]

Ranunculus sardous (Ranunculaceae)

Annual. Height Up to 40cm. **Habitat** Grassland and wasteland. **Distribution** All northern Europe except Ireland, Faeroe Is and Iceland. **Flowers** May–October. **Image size** Two-thirds life-size

Similar to the bulbous buttercup in flower size and with sepals that also point downwards, this species has distinctly paler blooms; they are best described as creamy rather than buttery. The foliage is similar to that of the bulbous buttercup, but is often shiny. The stems, sepals and leaves are all covered in long, fine hairs. Hairy buttercups are pollinated by both bees and flies.

Bulbous buttercup [5]

Ranunculus bulbosus (Ranunculaceae)

Perennial. Height Up to 40cm. **Habitat** Dry, grassy habitats. **Distribution** All northern Europe except Faeroe Is and Iceland. **Flowers** March–July. **In gardens** Fertile, moist but well-drained soil in full sun or partial shade. Borders, wild-flower and rock gardens. **Image size** Two-thirds life-size.

Along with meadow and creeping buttercups (pp.107 and 115), this is one of a trio often mistaken for each other. The bulbous buttercup's main distinguishing features – apart from liking drier soil – are the swollen stem-base (hence the name) and the way its sepals point down rather than hugging the petals. The typical bright yellow 20–30mm flowers grow on stout stems, and the three-lobed leaves are toothed, with the middle lobe suspended on a long stalk. This species attracts hordes of honeybees, so is a favourite of beekeepers and wildlife gardeners.

Oxlip [6]

Primula elatior (Primulaceae)

Perennial. Height 30cm. **Habitat** Damp shady habitats on clay soil. **Distribution** Northern Europe to southern Britain, Denmark and southern Sweden. **Flowers** April–May. **In gardens** Moist, humus-rich soil in full sun or partial shade. Borders, wild-flower meadows and woodland gardens. **Image size** Two-thirds life-size.

The stalked clusters of oxlip flowers are easily confused with the cowslip (p.84), both being arranged in one-sided racemes. But oxlip flowers are larger than cowslips, and are a delicate pale yellow instead of dark yellow. The flowers have no fragrance and are held slightly more upright than sweetly scented cowslip blooms. Nevertheless, it took botanists a long time to agree that the oxlip is a species in its own right rather than a hybrid between the cowslip and primrose (p.84). The clincher was the absence of primroses in an East Anglian location where oxlips grew in abundance.

YELLOW

Common bird's-foot-trefoil [1]

Lotus corniculatus (Fabaceae or Leguminosae)

Perennial. Height Up to 50cm. **Habitat** Grassland, heaths and roadsides. **Distribution** All northern Europe. **Flowers** June–September. **In gardens** Full sun. Rock gardens. **Image size** Two-thirds life-size.

The common name of this plant refers to the elongated seedpods, each with a hook at the tip, which resemble a bird's foot. In Britain, it has numerous other folk names, including eggs and bacon, devil's claws and granny's toenails. It is a variable plant. The flamboyant clusters of yellow flowers come in all shades to almost orange, and may be streaked with red. The leaves and stems may or may not be hairy and have three principal leaflets, plus an extra pair at the base of each petiole. The flowers attract bees, wasps and butterflies.

Horseshoe vetch [2]

Hippocrepis comosa (Fabaceae or Leguminosae)

Perennial. Height Up to 30cm. **Habitat** Dry grassland on chalk or limestone. **Distribution** Belgium, Britain, France, Holland and Germany. **Flowers** April–July. **In gardens** Poor, well-drained soil in full sun. Wild-flower and rock gardens, raised borders and troughs. **Image size** Two-thirds life-size.

The flowers of the horseshoe vetch can easily be confused with those of the bird's-foot trefoil. The pea-like blooms are bright yellow, occasionally streaked with red, and are held in clusters at the tip of each lanky stem. They have evolved to be pollinated by bumblebees and honey bees. It is the leaves and seedpods that set it apart. Each leaf consists of two rows of oval leaflets – up to eight pairs, with one more at the tip. The distinctive seedpods are curiously wavy and if broken into segments looks vaguely like a horseshoe. The botanical name comes directly from the Greek *hippos* (horse) and *krepis* (shoe). If the species is too vigorous, try the cultivar 'E.R. Janes'; it is less enthusiastic but thrives in paving cracks and wall crevices.

Tormentil [3]

Potentilla erecta (Rosaceae)

Creeping perennial. Height Up to 45cm. **Habitat** Grassland, hedgerows and roadsides. **Distribution** All northern Europe. **Flowers** June–September. **In gardens** Well-drained soil in full sun. Wild-flower gardens. **Image size** Two-thirds life-size.

Rather more creeping than erect, despite its botanical name, the tormentil produces flowers that are unusual among potentillas for having four petals instead of five. The glowing yellow blooms look a little like buttercups, and lure pollinating insects by exuding a potent nectar. Leaves of three to five toothed leaflets with silvery undersides look as if they have been threaded like beads at regular intervals along the supple stems. The dried roots have had a variety of uses from treating sunburn to oral infections and they produce a red dye.

Common cow wheat [4]

Melampyrum pratense (Scrophulariaceae)

Annual. Height Up to 60cm. **Habitat** Woodland, heaths and scrub. **Distribution** All northern Europe. **Flowers** May–October. **Image size** Life-size.

The genus name – from the Greek *melas* (black) and *puros* (wheat) – refers to the seeds, which look like black wheat seeds. Common cow wheat is a pretty plant, with pale buttery-yellow flowers of two lips, which are so tightly closed that only the hefty weight of a pollinating bumblebee can prise them apart. On each stem the flowers all face in the same direction. The foliage is very variable: long and slim or almost oval, toothed or not. It is a parasitic plant, taking moisture and nutrients from hosts such as eyebright (p.60).

Several myths surround it. Women who eat the seeds are supposedly more likely to have male babies, and cows who eat it are said to produce the best, deepest yellow butter. Make the most of its long flowering period in the wild-flower garden.

Perennial wall rocket [5]

Diplotaxis tenuifolia (Brassicaceae or Cruciferae)

Perennial. Height Up to 80cm. **Habitat** Rocky habitats, wasteland and walls. **Distribution** All northern Europe. **Flowers** May–September. **Image size** Two-thirds life-size.

Perennial wall rocket is very similar to the annual species (p.100), but for its larger flowers – up to 30mm across – which are held in dense clusters at the tip of the woody stems. The grey-green leaves, with toothed lobes, are arranged loosely around the base of the stem. The foliage smells strongly when crushed, but it is much less offensive than the sulphurous odour given off by the annual wall rocket. In Britain, it often grows along railways.

Greater celandine [6]

Chelidonium majus (Papaveraceae)

Tufted perennial. Height 40–90cm. **Habitat** Wasteland, walls, hedgerows and woodland. **Distribution** Northern Europe except Iceland. **Flowers** April–October. **In gardens** Any soil in full sun or partial shade. Shady borders, woodland and wild-flower gardens. **Image size** Half life-size.

Don't be fooled: this plant, a member of the poppy family, is not even remotely related to the lesser celandine (p.80), which is a relative of the buttercups. Not that its flowers are very poppy-like; they are less than 2.5cm across, with four rounded petals. They, and the long segmented seed capsules that follow, are more reminiscent of members of the cabbage family. The leaves consist of hairless, toothed leaflets rather like miniature oak leaves, with blue-green undersides. En masse, they form tall, conical mounds on brittle stems. Like the Welsh poppy (p.123), greater celandine has yellow sap rather than the white typical of true poppies. Dangerously poisonous and caustic, it can blister skin. The seeds are oily, attracting feeding ants who disperse them. Gardeners should beware the greater celandine's prolific self-seeding habits; weed out any unwanted seedlings.

Prickly sow thistle [1]

Sonchus asper (Asteraceae or Compositae)

Annual. Height Up to 1.5m. **Habitat** Wasteland, roadsides and cultivated land. **Distribution** Throughout northern Europe. **Flowers** June–August. **In gardens** A weed. **Image size** Half life-size.

The leaves of most sow thistles (see below and p.115) have soft prickly edges, but in this species they are exaggerated. The leaves clasp the stem with distinctive rounded lobes and look more aggressive than they really are – in fact, the leaves were regularly eaten as a salad vegetable. The flowers are very like those of the dandelion, a close relation. Flat luminous-yellow discs of florets hold their faces skywards in loose clusters after unfurling from juicy, tight buds; they are followed by dandelion-like seed heads. Each seed is carried on the wind by a tiny parachute of fine hairs, so the plant can easily colonise garden soil; it has a long taproot, which needs to be dug out with a spade. The stems and roots are filled with a thick white sap, which was thought to encourage a mother's milk flow.

Smooth sow thistle [2]

Sonchus oleraceus (Asteraceae or Compositae)

Annual. Height Up to 1.5m. **Habitat** Waste and burned land, roadsides and farmland. **Distribution** All northern Europe. **Flowers** June–August. **In gardens** A weed. **Image size** Half life-size.

Smooth sow thistle hardly differs from its prickly relation (see above), except that the foliage is darker and less glossy, and has less prickly margins. Instead of clasping the stem with rounded lobes, the leaves of the smooth sow thistle have more triangular-shaped lobes. Like all sow thistles, it exudes a thick, white sap. It was thought that pigs ate the plant – supposedly to increase their milk flow after giving birth – hence its British common name. Boiled and smothered in butter, it tastes like spinach; or it can be eaten raw in salads.

Henbane [3]

Hyoscyamus niger (Solanaceae)

Annual or biennial. Height Up to 80cm. **Habitat** Sand, shingle and wasteland. **Distribution** All northern Europe but rare in Ireland. **Flowers** May–September. **Image size** Two-thirds life-size.

Henbane is a malevolent plant, and the pale yellow flowers, with their net of purple veins and purple anthers, have a suitably sinister appearance and noxious odour. Trumpet-shaped and up to 30mm long, they nestle in one-sided spikes among clumps of oval, lobed or coarsely toothed leaves. These and the stem have coarse sticky hairs. The seeds are released through a lid at the top of the seed capsule.

All parts of henbane are poisonous. It causes hallucinations, and in the Middle Ages was an essential ingredient in witches' brews. The narcotic properties of henbane were used for centuries to cure toothache. Even today, drugs extracted from it are used in tiny doses to treat conditions ranging from sea-sickness to mental illness.

Yellow corydalis [4]

Corydalis lutea (syn. *Pseudofumaria lutea*; Fumariaceae)

Perennial. Height 30cm. **Habitat** Walls and stony habitats. **Distribution** Introduced from Alps; naturalised in northern Europe except Scandinavia. **Flowers** May–October. **In gardens** Well-drained soil in full sun or partial shade. Rock gardens, woodland gardens and mixed borders. **Image size** Two-thirds life-size.

The delicate ferny foliage of yellow corydalis forms airy mounds through which the slim flower stems emerge in spring. At their tip, the racemes of up to 16 dainty sulphur-yellow flowers, 12–20mm long, bloom until early autumn. Yet, despite appearances, the plant thrives in the least hospitable of environments, thanks to its penetrating roots. It is probably better known as a garden plant than a wild flower, and is good value for money, self-seeding wherever it puts down roots.

Spotted medick [5]

Medicago arabica (Fabaceae or Leguminosae)

Annual. Height Up to 60cm. **Habitat** Grassland and wasteland. **Distribution** Britain, Belgium, France, Germany and Holland; naturalised in Ireland and southern Sweden. **Flowers** April–August. **Image size** Two-thirds life-size.

The simple bright yellow pea-like flowers of the spotted medick give no indication of the unique seedpods to come. The tightly coiled pods are armed with a layer of spines, so they can hitch a ride on the fur of passing animals. Only one seed of the several in each pod will germinate successfully and, instead of the pod splitting open, a single young root emerges through the wall of the pod. The distinctly heart-shaped and ribbed leaves are decorated with dark spots at their centre; these were thought to resemble drops of Christ's blood that fell from the cross on to clover below.

Yellow archangel [6]

Lamiastrum galeobdolon (syn. *Lamium galeobdolon*; Lamiaceae or Labiatae)

Perennial. Height 30–60cm. **Habitat** Shady habitats on chalky soil. **Distribution** Britain, north to southern Sweden; naturalised in Norway and Finland. **Flowers** April–July. **In gardens** Moist, well-drained soil in full or partial shade. Ground cover, but extremely invasive. **Image size** Two-thirds life-size.

Closely related to white deadnettle (p.32), this species prefers the seclusion of woodland and hedge-banks rather than the open roadsides favoured by its cousin. In Britain it is sometimes known as weasel-snout. This refers to the 20mm flowers, which consist of a furry upper hooded lip – supposedly like a weasel's head – and a bright yellow lower lip of three lobes streaked with red. They are arranged in dense whorls on the stout square stems. The dark green leaves are arranged in pairs at the base of each flower whorl. Given a chance in the garden, yellow archangel will spread far and wide on its creeping rootstock.

Kidney vetch [2]

Anthyllis vulneraria (Fabaceae or Leguminosae)

Perennial. Height 60cm. **Habitat** Grassland on chalky soil, especially near the sea. **Distribution** All northern Europe. **Flowers** June–September. **In gardens** Well-drained soil in full sun. Rock gardens and wild-flower meadows. **Image size** Life-size.

It's quite common to find the flower heads of kidney vetch in a state of flux, one half having its small (12mm) pea flowers fully open and the other in bud or fruit. The kidney shape of the flower heads was thought to indicate the plant's properties as a cure-all for kidney complaints. It was also regularly used to heal wounds. The species name *vulneraria* comes from the Latin for 'wound'. The base of each flower head is cushioned by finger-like bracts, which are covered in woolly hairs, as are the numerous narrow, oval leaflets and rather lax stems. Kidney vetch is so common and widespread that there are bound to be a few variations, including not only yellow flowers but orange, red, purple or white – all of them pollinated by bumblebees.

Rough hawkbit [3]

Leontodon hispidus (Asteraceae or Compositae)

Perennial. Height Up to 60cm. **Habitat** Grassy habitats on well-drained, chalky soil. **Distribution** Throughout northern Europe. **Flowers** June–October. **Image size** Life-size.

The solitary bright yellow or almost orange 25–40mm flower heads of rough hawkbit – not unlike a dandelion's – appear at the tip of simple, unbranched stems. Before the flowers unfurl, the buds are encased in protective bracts covered in white bristles (*hispidus* means 'bristly'). At the base of each stem there is a relaxed rosette of oblong leaves, with deeply toothed margins – though less deeply than a dandelion's. The whole plant is hairy. The seeds are wind-borne, carried along by a parachute of fine hairs, which are particularly feathery.

Crosswort [1]

Cruciata laevipes (Rubiaceae)

Creeping perennial. Height 30–60cm. **Habitat** Grassland, hedgerows and verges on chalky soil. **Distribution** Britain, except extreme north of Scotland, north to Holland and Germany. Introduced in Ireland. **Flowers** April–June. **Image size** Life-size.

Crosswort has extremely hairy yellow-green square stems. The egg-shaped leaves are arranged in whorls of four up the stem, with neat posies of tiny yellow star-like flowers in the angle between stem and leaf. The flowers, which are followed by dark purple berries, exude a tantalising scent of honey that must be irresistible to pollinating insects. The 16th-century herbalist John Gerard is credited with first finding crosswort in Britain, at Hampstead, then just north of London.

Common evening primrose [4]

Oenothera biennis (Onagraceae)

Annual or biennial. Height Up to 1.5m. **Habitat** Sand dunes and wasteland. **Distribution** Naturalised in most of northern Europe except Finland, Iceland and Faeroe Is. **Flowers** June–September. **In gardens** Well-drained soil in full sun. Herbaceous and mixed borders. **Image size** Two-thirds life-size.

The evening primrose was brought from North America in the 17th century, first landing in Italy before spreading across the rest of Europe. Although it grows in the wild, it is also well known in gardens. The blooms – deflated balloons by day – are transformed into voluptuous, shimmering yellow goblets at dusk, ready for pollination by nocturnal insects. The wavy-edged leaves have red veins and appear all the way up stems that increase in height throughout the season. By September, as the days shorten, the flowers stay open all day in a last-ditch attempt to attract insects. Evening primrose oil, extracted from the seeds, has been used to treat pre-menstrual tension, eczema and even hangovers.

Goat's-beard, Jack-go-to-bed-at-noon [5]

Tragopogon pratensis (Asteraceae or Compositae)

Annual or perennial. Height Up to 75cm. **Habitat** Grassland; occasionally sand dunes. **Distribution** All northern Europe. **Flowers** June–July. **Image size** Two-thirds life-size.

Goat's-beard is named for its voluminous seed heads, like overblown dandelion 'clocks'. Each sturdy stem carries a solitary, bright yellow 18–40mm flower head, which closes by midday – hence its second, country name. Both the long taproot and the spindly, grass-like leaves are edible; the taproot was cooked and eaten with butter like parsnips, while the leaves can be eaten raw in salads. In France, goat's-beard is called *barbe à Jean* (St John's beard), and is picked to celebrate St John's Eve.

Common cat's-ear [6]

Hypochaeris radicata (Asteraceae or Compositae)

Perennial. Height Up to 60cm. **Habitat** Grassland. **Distribution** All northern Europe except Finland. **Flowers** June–September. **In gardens** Wild-flower gardens on well-drained soil; also a persistent lawn weed. **Image size** Two-thirds life-size.

This is another plant that, to the untrained eye, resembles dandelion and hawkweed (pp.84 and 99). It is distinguished by the tiny scales, which appear along the upright flower stems and are thought to resemble cat's ears. Without this, the luminous yellow flowers and rosette of hairy leaves, with their triangular teeth, could be mistaken for those of the dandelion. Cat's-ear is a stubborn irritant, having a long taproot that makes it difficult to eradicate, but it is a welcome addition to wild-flower meadows.

Bristly oxtongue [1]

Picris echioides (Asteraceae or Compositae)

Annual or biennial. Height Up to 80cm. **Habitat** Disturbed ground, wasteland and streamsides. **Distribution** Southern Britain and Ireland to Holland and Germany. **Flowers** June–November. **Image size** Half life-size.

This plant is not called bristly (or prickly) oxtongue for nothing. All parts, except the flower heads, are smothered in rough bristles with hooked ends. The deep yellow dandelion-like flowers have strap-shaped rays and, once pollinated, turn into spherical, dandelion-clock seed heads. The wavy-edged leaves are coarse and covered in pimples. Even so they were once used to make pickles and are supposed to have a decent flavour as long as they are boiled. The name *Picris* (from the Greek for 'bitter') alludes to the thick white bitter sap.

Honeysuckle, woodbine [2]

Lonicera periclymenum (Caprifoliaceae)

Twining climber. Height Up to 7m. **Habitat** Hedges, woodland and rocks. **Distribution** All northern Europe as far north as central Scandinavia. **Flowers** June–October. **In gardens** Humus-rich, well-drained soil in full sun or partial shade. Against walls and fences, or through mature trees. **Image size** Half life-size.

Honeysuckle's deliciously spicy-sweet perfume is redolent of warm summer evenings. This romantic climber clambers through native hedgerows on stems that turn woody and silvery-grey as they mature. The 5cm-long tubular flowers are held in clusters; they start creamy-white but turn pale yellow once pollinated and may be flushed with purple. Their scent is strongest at dusk, to attract pollinating moths who can detect it hundreds of metres away. By day honeysuckle is pollinated by bees who have to delve deep into the flower's narrow tube to extract nectar. Children have for centuries simply sucked the nectar from the narrow end. Clusters of glossy scarlet berries steal the show in the autumn.

Small balsam [3]

Impatiens parviflora (Balsaminaceae)

Annual. Height 20–60cm. **Habitat** Damp woodland, wasteland and hedgerows. **Distribution** Naturalised in most of northern Europe. **Flowers** July–November. **Image size** Two-thirds life-size.

Small balsam was introduced from its native Russia in the mid-19th century, and soon made itself at home. It has dainty, pale yellow lipped flowers typical of *Impatiens* species, with five petals and a spur where nectar is stored. At only 18mm across, they are smaller than those of many other balsams. The pointed, oval leaves have toothed edges and are arranged alternately up the stem. Balsams have a dramatic method of releasing their seed. As soon as the club-shaped seed capsule is ripe, it explodes noisily at the slightest touch, catapulting seeds far and wide.

Wild mignonette [4]

Reseda lutea (Resedaceae)

Bushy perennial/biennial. Height 75cm. **Habitat** Waste and arable land on chalky soil. **Distribution** Europe as far north as southern Scandinavia. **Flowers** June–September. **Image size** One-third life-size.

Wild mignonette is related to weld (p.120), a dye plant, and common mignonette (*Reseda odorata*), which is grown in gardens for its sweet fragrance. Sadly, wild mignonette possesses neither of these attributes, but it is still a pretty plant with racemes of slightly straggly pale yellow flowers on stiff, upright stems. They are pollinated by bees and are followed by funnel-like seed capsules. The leaves are deeply lobed and pinnate, with three finger-like leaflets.

Creeping cinquefoil [5]

Potentilla reptans (Rosaceae)

Perennial. Height Up to 10cm; stems up to 1m long. **Habitat** Grassland, wasteland and road verges. **Distribution** All northern Europe except Faeroe Is and Iceland. **Flowers** June–September. **In gardens** A weed. **Image size** Two-thirds life-size.

Despite its pretty buttercup-like flowers, in the wrong place, creeping cinquefoil is a tenacious weed. The creeping roots throw up long, stringy runners, which are pinned to the soil by clusters of fine roots at each leaf node. Try using a dandelion weeder to remove it from lawns. The leaves each have five toothed fingers, and at each node a long flower stem emerges, with a single yellow five-petalled flower, up to 25mm across, at its tip. In the Middle Ages creeping cinquefoil was a favourite ingredient of magic spells – including the gruesome Witches' Ointment, which included the fat of dead children.

Biting stonecrop [6]

Sedum acre (Crassulaceae)

Evergreen perennial. Height 10cm. **Habitat** Heaths, sand dunes and rocky habitats. **Distribution** All northern Europe. **Flowers** May–July. **In gardens** Dry soil in full sun. Rock gardens and walls. **Image size** Two-thirds life-size.

Above mats of chubby, succulent foliage, star-shaped, bright yellow flowers are produced on short stems all summer long. Anchored by fine fibrous roots, biting stonecrop grows where most plants cannot get a foothold, such as rock crevices. The young foliage is tinged with crimson, but the leaves have a pungent odour and acrid taste – a warning that every part of the plant causes stomach upsets if eaten. Avoid the sap too, which will irritate sensitive skin. In spite of all this, biting stonecrop has been used as a remedy for scurvy. For rock or coastal gardens, *S. acre* 'Aureum' is a bright cultivar, with flowers and leaves both a luminous yellow.

Tall melilot [1]

Melilotus altissimus (Fabaceae or Leguminosae)

Biennial or perennial. Height Up to 1.5m.
Habitat Waste and rough ground. **Distribution** Most of northern Europe except Faeroe Is and Iceland; in many areas naturalised. **Flowers** June–August. **Image size** Two-thirds life-size.

Melilots have been cultivated since at least the 16th century. Herbalists grew them first for the dried leaves and flowers, especially for anti-inflammatories. They have also been grown for bees, who are attracted to their nectar. As a result meliots have become widely naturalised. The flowers are perfectly constructed for pollination; when a heavy insect lands on the lower petals, the stigma and anthers are pushed against its body, collecting and depositing pollen. Tall melilot can be distinguished by its downy seedpods, which have net-like markings.

Lesser spearwort [2]

Ranunculus flammula (Ranunculaceae)

Perennial. Height Up to 50cm. **Habitat** Wet habitats. **Distribution** All northern Europe. **Flowers** May–September. **In gardens** Still or slow-moving water in full sun. Bog gardens and pond margins (but poisonous). **Image size** Two-thirds life-size.

Although it looks innocent enough, the lesser spearwort is, in fact, viciously poisonous, with the capacity to kill animals as large as sheep and cattle. It is the most poisonous of its genus except for the celery-leaved crowfoot (p.115). The abundant flowers, which appear on multi-branched stems, may look like those of a buttercup, but the leaves are smooth and spear-shaped rather than lobed and divided – the crucial clue that exposes lesser spearwort for what it is. The sap can cause blisters if it comes into contact with skin – a possible explanation for the species name *flammula*, which means 'inflammatory', and a side-effect that beggars once exploited in a mischievous ruse to gain sympathy.

Meadow vetchling [3]

Lathyrus pratensis (Fabaceae or Leguminosae)

Climbing perennial. Height Up to 1.2m. **Habitat** Grassy habitats, hedgerows and roadsides. **Distribution** All except far northern Europe. **Flowers** May–August. **In gardens** Fertile, well-drained soil in full sun. Wild-flower gardens. **Image size** Two-thirds life-size.

The pretty meadow vetchling is closely related to the everlasting peas (pp.155 and 163), as well as to garden sweet peas, and looks very similar except that it has deep yellow flowers. Unlike sweet peas, however, the stems of meadow vetchling are wingless. It has up to 12 flowers in each cluster; they are pollinated by bees who stretch their long tongue to the limit to extract nectar from the depths of the flower tube. Pairs of oval leaflets clasp the stem and, at the centre of each pair, tendrils grow in all directions to support the plant as it clambers up any available support. The meadow vetchling's root nodules contain bacteria that enrich the soil with nitrogen.

Herb bennet, wood avens [4]

Geum urbanum (Rosaceae)

Perennial. Height Up to 70cm. **Habitat** Woods and hedgerows. **Distribution** All northern Europe except far north. **Flowers** May–September. **In gardens** Moist but well-drained soil in sun or partial shade. Wild gardens. **Image size** Two-thirds life-size.

The unassuming, upturned 15mm flowers of herb bennet give no hint of its supposed powers over evil. The clove-scented roots gave it the German name *Nelkenwurz* (clove root), and in the 15th century people kept the roots in the house to ward off the devil. The English name comes from the Latin *herba benedicta*, or 'blessed herb', as does the French *herbe de Saint-Benoit*.

The leaves consist of three main toothed leaflets and a pair of smaller stipules at their base. The hairy stems are brittle, so it can be tough to remove, as the stems tend to snap, leaving the roots behind.

Lesser meadow rue [5]

Thalictrum minus (Ranunculaceae)

Perennial. Height 50–120cm. **Habitat** Grassland, sand dunes and rocky places; also near streams. **Distribution** Most of northern Europe except Faeroe Is and Iceland. **Flowers** June–August. **In gardens** Well-drained soil in sun; may become invasive in rich soil. Wild-flower gardens. **Image size** Two-thirds life-size.

Surprisingly, meadow rues are close relations of buttercups, although it is not immediately apparent. Instead of golden cup-like flowers, those of the wiry lesser meadow rue consist mostly of straggly greenish-yellow stamens on pendulous stalks, which give them a rather droopy, shaggy appearance. Each lacy leaf consists of three blue-grey lobed leaflets on slender stems. It is a variable species, growing much taller in damp conditions. Lesser meadow rue is mostly wind-pollinated.

Hawkweed [6]

Hieracium agg. (Asteraceae or Compositae)

Perennial. Height Up to 80cm. **Habitat** Grassland, open woodland, heaths, banks, cliff-tops and rocky habitats. **Distribution** All northern Europe except Iceland and far north. **Flowers** May or June onwards. **In gardens** A weed. **Image size** Two-thirds life-size.

There are countless very similar hawkweed species, often simply grouped as *Hieracium* agg. (or aggregate). These arose because, although hawkweeds reproduce without pollination, the seedlings are not identical clones of their single parent, but vary slightly. So, although most hawkweeds have yellow flowers in clusters on branched stems, some are red or golden. The ray florets have toothed tips and the bracts are usually substantial and hairy, making the flower heads look as if they are held by a tight corset. The stem leaves are alternate and usually hairy; there is also usually a rosette of basal leaves. The seed heads are soft and downy, each seed having a feathery parachute.

Creeping yellow-cress [1]

Rorippa sylvestris (Brassicaceae or Cruciferae)

Creeping perennial. Height Up to 60cm. **Habitat** Damp habitats and disturbed ground. **Distribution** Most of northern Europe except Faeroe Is and northern Scandinavia. **Flowers** June–September. **Image size** Half life-size.

Wherever celery-leaved crowfoot (p.115) grows, creeping yellow-cress is likely to be close by. They enjoy similar environments where there is still or slow-moving water. Creeping yellow-cress has angled, branched stems but tends to have a straggly habit. The bright yellow flowers, with four petals, appear in sparse but broad sprays. They are pollinated by bees and flies, and are followed by long slim seedpods, up to 18mm long, with two rows of seeds. The spear-shaped leaves are deeply toothed and arranged spirally up the stem.

Annual wall rocket [2]

Diplotaxis muralis (Brassicaceae or Cruciferae)

Annual or biennial. Height Up to 60cm. **Habitat** Rocks, walls, wasteland and arable land. **Distribution** Naturalised in Britain, Holland, Sweden, France and southern Germany. **Flowers** June–September. **Image size** Half life-size.

The alternative name stinkweed is more apt for this plant, given the noxious smell that the stems give off when crushed – it is like foxes or sulphurous stink bombs, depending on your sense of smell. While the stench deters grazing animals, the surprisingly pleasant fragrance of the flowers attracts pollinating bees and flies. The mustard-like blooms each have four rounded petals and are arranged alternately up the stem. The flowers are followed by long, slim seedpods typical of the brassica family, which contain two rows of rounded seeds. Rosettes of lobed leaves form at the base of the plant.

Common ragwort [3]

Senecio jacobaea (Asteraceae or Compositae)

Biennial or perennial. Height Up to 1.5m. **Habitat** A weed of grassland, wasteland, dunes and roadsides. **Distribution** Most of northern Europe. **Flowers** June–November. **Image size** One-quarter life-size.

Golden, daisy-like common ragwort flowers, growing en masse across pasture, make a splendid high-summer spectacle. But it's not one that any horse-keeper or livestock farmer will welcome. The plant contains alkaloid poisons that can cause fatal liver damage in cattle and horses, and in Britain it must, by law, be removed from land where livestock graze. As a result, common ragwort is most often seen on fallow or set-aside fields. Yet, ironically, a ragwort infusion was once thought to cure staggers, an equine brain disease. Sheep seem to be more or less immune to it.

Common ragwort is a variable plant: hairy or hairless, perennial or biennial, and of varying height. The flowers appear in branched clusters; by the time they are fully open, the lobed leaves are withering.

Common rock-rose [4]

Helianthemum nummularium (Cistaceae)

Evergreen sub-shrub. Height Up to 50cm. **Habitat** Rocky places and dry grassland on chalky soil. **Distribution** Most of northern Europe except Faeroe Is, Iceland and Norway. **Flowers** June–September. **In gardens** Fertile, well-drained soil in full sun. Rock gardens and ground cover. **Image size** Two-thirds life-size.

Rock-garden aficionados will already know the sprawling wiry stems of the golden-yellow rock-rose and its fragile, papery blooms. The pairs of elliptical leaves, which are well adapted for dry conditions, are dark green and leathery above but white and hairy underneath. *Helianthemum* means 'sun-flower', and the 20mm flowers, which are teeming with pollen, open fully only when the sun is out. On grey days and at night, the rock-rose's petals fold up.

Narrow-leaved bird's-foot-trefoil [5]

Lotus glaber (syn. *L. tenuis*; Fabaceae or Leguminosae)

Sprawling perennial. Height Up to 90cm. **Habitat** Grassland and coastal habitats on chalky soil. **Distribution** All northern Europe except northern Scandinavia. **Flowers** June–August. **Image size** Two-thirds life-size.

Although very similar to the common bird's-foot trefoil (p.91), the foliage sets this species apart. As with other trefoils, each leaf consists of three leaflets, but here they are noticeably long and slender. Clusters of two to four pure yellow pea flowers give way to long seedpods – also pea-like, and up to 3cm long – which are shaped like a bird's toe and claw. They twist to release their seed when ripe.

Black medick [6]

Medicago lupulina (Fabaceae or Leguminosae)

Scrambling or prostrate annual. Height Up to 80cm. **Habitat** Grassy habitats. **Distribution** All northern Europe. **Flowers** April–September. **Image size** Two-thirds life-size.

The British common name sounds as if it has been lifted from the pages of an ancient book of spells or herbal remedies. Yet black medick is, in fact, named after the Medes, a race of ancient Persian peoples. The name was originally given by the Romans to the closely related lucerne (p.182).

If you look closely, you can see that the fluffy yellow flower heads, which emerge on slender stems at the leaf nodes, are crammed with up to 50 individual flowers. They are followed by unique bean-like seedpods, which coil up and turn black. Black medick is often mistaken for the hop trefoil (p.112), and the name *lupulina* means 'hop-like'. But, unlike the hop trefoil, its seedpods don't remain encased in the dried flower heads and they also have prominent veins along their length. In Ireland, black medick is one of several plants that masquerade as the shamrock, and its leaves are sold as such on St Patrick's Day.

Agrimony [1]

Agrimonia eupatoria (Rosaceae)

Perennial. Height 10–60cm. **Habitat** Dry grassland, often on chalk. **Distribution** Northern Europe except far north. **Flowers** June–August. **Image size** Two-thirds life-size.

Spikes of pale yellow, apricot-scented 8mm flowers, each with five petals, rise from rosettes of leathery, toothed leaves. They are followed by hooked, burr-like seed heads, helping dispersal as they attach to passing animals. Agrimony is an ancient remedy for all sorts of ailments. The 1st-century physician Dioscorides used it as a remedy for snake bites, dysentery and liver disease, and in the 16th century, John Gerard, believed that 'a decoction of the leaves is good for them that have naughty livers.' It is still used in preparations for catarrh and digestive disorders.

Yellow-horned poppy [2]

Glaucium flavum (Papaveraceae)

Biennial or short-lived perennial. Height 30–90cm. **Habitat** Sandy coastal habitats and wasteland. **Distribution** Northern Europe to central Scotland and southern Norway. **Flowers** June–September. **In gardens** Well-drained soil in full sun. Borders and gravel gardens. **Image size** Two-thirds life-size.

The sturdy stems and rosettes of hairy silvery-grey leaves equip the yellow-horned poppy to withstand harsh coastal conditions, but they belie the papery fragility of its solitary, golden-yellow flowers. These unfurl with great delicacy from a hairy protective casing. As each flower dies, a long, slim seed capsule – the so-called horn – takes its place; it splits lengthways to release the seed. The entire plant is poisonous, and Geoffrey Grigson in *The Englishman's Herbal* tells a nasty 17th-century tale of how the roots, mistaken for those of the sea holly, were cooked in a pie, causing severe delirium.

Corn marigold, gold [3]

Chrysanthemum segetum (Asteraceae or Compositae)

Perennial or annual. Height Up to 80cm. **Habitat** Wasteland, disturbed ground and arable fields. **Distribution** Northern Europe except N. Scandinavia, Iceland. **Flowers** June–August. **In gardens** Well-drained soil in sun. Wild-flower gardens and annual borders. **Image size** Two-thirds life-size.

The upturned flat golden daisies, 50mm across, may look stunning en masse, but the corn marigold has long been the bane of arable farmers. It was described as an 'eyll wede [that] groweth commonlye in barleye and pees', and in the 12th century Henry II of England issued a decree ordering its destruction. Yet seedlings by the thousand emerge whenever farmland is disturbed, suggesting that the seeds can lie dormant for many years, waiting to germinate. Despite its persistent nature, it is a charming plant with an attractive scent and is sometimes grown as a garden annual or perennial. Also called simply 'gold', it has influenced many British place names, such as Goldhanger in Essex and Golding, Shropshire. The sharply toothed, fleshy leaves clasp the sturdy stems.

Yellow flag iris [4]

Iris pseudacorus (Iridaceae)

Rhizomatous perennial. Height 1–1.5m. **Habitat** Marshes and freshwater margins. **Distribution** All northern Europe except Iceland. **Flowers** June–August. **In gardens** Wet or moist soil in full sun. Pool margins, bog gardens and moist borders. **Image size** Two-thirds life-size.

A common sight at the edge of ponds, lakes and rivers, yellow flag irises are also handsome marginal plants for garden pools. Each tall, sturdy stem produces up to three deep yellow blooms; sword-like leaves rise from the base. The Frankish King Clovis first wore the flower as a heraldic device in the 5th century; later, under Louis VII, it was the basis of the *fleur-de-lys* emblem.

Adapted for pollination by bees and hover flies, the flowers protect their cache of pollen from rain with umbrella-like stigmas that arch over the stamens. There follow green seed capsules like miniature cucumbers, which split when ripe to release the smooth, flat seeds. The boiled roots were sometimes used for soothing bruises and cramps and were powdered to make snuff. Garden cultivars include 'Alba', with pale cream flowers; dark yellow 'Golden Fleece'; and 'Variegata', whose leaves have striking white stripes.

Moth mullein [5]

Verbascum blattaria (Scrophulariaceae)

Biennial. Height 10–60cm. **Habitat** Damp habitats and wasteland. **Distribution** Belgium, Holland, France and Germany, rare in Britain. **Flowers** June–October. **Image size** Two-thirds life-size.

The moth mullein, so named because it is pollinated by moths, is easily distinguished from other species of *Verbascum* by the arrangement of the flowers; instead of appearing as crowded spikes, they are sparse and loosely arranged along upright stems. At the base of the stem, the edges of the lanceolate leaves have blunt teeth, but towards the stem tip they are smaller and more triangular in shape, and hug the stem. This arrangement allows the smaller leaves to direct rainwater on to the larger ones at the base and thus on to the roots.

Common cudweed [1]

Filago vulgaris (Asteraceae or Compositae)

Annual. Height Up to 40cm. **Habitat** Bare sandy soil and heathland. **Distribution** All northern Europe. **Flowers** July–August. **Image size** Life-size.

Common cudweed is one of the more unusual members of the daisy family. Woolly, lance-shaped leaves obscure the upright stems, which are branched at the tips where clusters of pale creamy-yellow flowers nestle. In the 17th century, farmers fed it to cattle that were unable to produce semi-digested cud, which they usually regurgitate and chew again. Cudweed could prompt the process to restart. The herbalist Nicholas Culpeper used the leaves to heal wounds. The genus name *Filago* comes from the Latin *filium* (thread), referring to the plant's woolly texture.

Rock samphire [2]

Crithmum maritimum (Apiaceae or Umbelliferae)

Perennial. Height 20–45cm. **Habitat** Cliffs, coastal rocks, sand and shingle. **Distribution** Coasts from France to Holland and parts of Britain and Ireland. **Flowers** June–October. **Image size** One-sixth life-size.

Rock samphire was so much in demand in the 16th and 17th centuries that collectors risked their necks to scale the sheer cliff faces where it grew. It was notorious as a trade, and Shakespeare alluded to the risks in *King Lear*. The fleshy stems and leaves were – and still are – cooked like asparagus, and the leaves pickled or made into sauces. If eating rock samphire takes your fancy, preserve wild populations by cultivating plants bought from specialist nurseries.

The plant's succulent, ridged, branched stems and smooth, untoothed leaves allow rock samphire to survive harsh, drying coastal conditions; the leaves smell sulphurous when crushed. The greenish-yellow flowers are held in umbels, which can have up to 36 rays. The name comes from the French *herbe de St-Pierre*.

Yellow-wort [3]

Blackstonia perfoliata (Gentianaceae)

Annual. Height 20–50cm. **Habitat** Grassland on chalky soil, and dunes. **Distribution** Belgium, Britain, France, Germany, Holland and Ireland. **Flowers** June–October. **Image size** Half life-size.

Yellow-wort is a slender yet sturdy plant, with grey-green stems and leaves that have a smooth rubbery texture, helping the plant to tolerate dry conditions. The pretty, bright yellow flowers – up to 15mm across and held in panicles on branched stems – open only in full sun. They usually have eight petals, unlike most other members of the gentian family, whose flower parts tend to be in fours and fives. There is a rosette of oval leaves at the base, but pairs of leaves grow along the stem; they are fused together at the base, so they look as if they are threaded on the stem – hence the species name *perfoliata*, ('through-leaved'). Yellow-wort was once used to make a yellow dye and, although it had medicinal uses, it is very bitter-tasting.

Greater spearwort [4]

Ranunculus lingua (Ranunculaceae)

Aquatic perennial. Height Up to 1.2m. **Habitat** Marshes and pond margins. **Distribution** Most of northern Europe except Faeroe Is and Iceland. **Flowers** June–September. **In gardens** Still or slow-moving water in mud, full sun. Pool margins, bog gardens and streamsides. **Image size** Half life-size.

It may not be as common as lesser spearwort (p.99), but this taller relative is very distinctive with its majestic stems and obviously spear-shaped foliage. The stems are hollow and smooth instead of grooved like its relative. The golden goblet flowers are very similar to the lesser spearwort's but are more than twice as big, up to 5cm across. Take care not to plant greater spearwort in small pools; it is so vigorous that you will have more plant than pond before you know it.

Lady's bedstraw [5]

Galium verum (Rubiaceae)

Sprawling perennial. Height Up to 1m. **Habitat** Dry grassland on chalky soil. **Distribution** All northern Europe except Faeroe Is. **Flowers** June–September. **Image size** Half life-size.

Traditionally, sheaves of lady's bedstraw were used to stuff mattresses. The name comes from legends of the Nativity dating back to the Middle Ages. Mary supposedly lay on a mattress of bedstraw to give birth to Jesus and in Germany the plant is called *Marienbettstroh* (Mary's bedstraw). The yellow, four-petalled tiny flowers are honey-scented, but they smell distinctly hay-like when they dry. The flowers are arranged in bushy panicles on branched stems, which emerge from the leaf nodes; they brim with nectar and attract hordes of butterflies. The elegant whorls of long, narrow leaves appear along the length of the stems. The flowers contain an astringent that curdles milk and was used in cheese-making; the Danish name *melklobe* echoes this.

Wild parsnip [6]

Pastinaca sativa (Apiaceae or Umbelliferae)

Biennial. Height Up to 1.8m. **Habitat** Grassy habitats and roadsides on chalky soil. **Distribution** Belgium, Britain, Holland, France, Germany and Ireland. **Flowers** July–August. **Image size** Half life-size.

This ancestor of our domestic culinary vegetable was eaten long ago with salted fish, and used to make sauces and cakes; in fact, the name comes from the Latin *pastus*, meaning 'food'. In Ireland it was used for brewing beer. A 19th-century experiment proved that, planted in rich garden soil, the wild parsnip would produce a more substantial, tasty root. Its neat clusters of tiny, yellow flowers are held in umbels on stiff, grooved, usually hollow stems. The coarse, toothed leaflets, each on their own stem, are covered in hairs. The sap of the wild parsnip can blister skin.

Common or yellow toadflax [1]

Linaria vulgaris (Scrophulariaceae)

Perennial. Height 40–80cm. **Habitat** Grassland, roadsides and wasteland. **Distribution** All northern Europe except Faroe Is and Iceland. **Flowers** July–October. **In gardens** Light, sandy soil in full sun. Borders and gravel gardens, but very invasive and difficult to eradicate. **Image size** Two-thirds life-size.

The vibrant yellow snapdragon-like flowers of common toadflax are an attractive summer sight. The way they open when squeezed at the sides has led to a some unflattering colloquial names – not least the Norwegian *torkesmund* (cod's-mouth) and the British squeeze-jaw – as well as the more endearing bunny-mouth. They are up to 30mm long and form spikes on stiff, upright stems; they have two orange-spotted lower lips and a long, pointed spur where nectar is stored. Only long-tongued bees are heavy enough to press down on the lower lip and extract the nectar. The long, narrow leaves are arranged in spirals along the stems and look very flax-like.

Rock stonecrop [2]

Sedum forsterianum (Crassulaceae)

Evergreen perennial. Height 35cm. **Habitat** Rocks, screes and rocky woodland. **Distribution** North-western Europe as far north as south Scandinavia, rare in Ireland. **Flowers** June–July. **In gardens** Well-drained soil in full sun. Rock and gravel gardens. **Image size** Two-thirds life-size.

The mat-forming rock stonecrop is often mistaken for its close relative the reflexed stonecrop (p.112). It produces the same crowded clusters of starry flowers, which are bright or pale yellow, at the tip of fleshy, pale pink stems enveloped in grey, succulent, pointy leaves. The most obvious difference is that the non-flowering stems have distinctly bushy clusters of foliage at the tip and any dead leaves persist, clinging on underneath – reflexed stonecrop drops its dead leaves. Given space, rock stonecrop will happily thrive in a rock garden.

Alpine lady's mantle [3]

Alchemilla alpina (Rosaceae)

Creeping perennial. Height 10–20cm. **Habitat** Mountain habitats and some grassy places. **Distribution** Most of northern Europe except Belgium, Holland and Denmark, rare in Ireland. **Flowers** June–August. **In gardens** Most garden soils in sun or partial shade. Borders or rock gardens. **Image size** Two-thirds life-size.

Like a smaller version of the common lady's mantle (p.221), alpine lady's mantle is a less sprawling and more compact plant in every way. The leaves consist of five to seven finger-like lobes, obviously toothed and each with a smooth top surface and undersides covered in fine silver hairs. These help to slow the evaporation of moisture reserves – vital in harsh mountainous environments. Instead of the billowing clouds of flowers produced by its larger cousin, alpine lady's mantle carries its tiny greenish-yellow blooms in small, pert clusters.

Creeping Jenny [4]

Lysimachia nummularia (Primulaceae)

Evergreen, creeping perennial. Height Up to 60cm. **Habitat** Damp habitats. **Distribution** Most of northern Europe except northern Scotland and Iceland. **Flowers** May–July. **In gardens** Moist, well-drained soil in full sun or partial shade. Ground cover. **Image size** Two-thirds life-size.

Creeping Jenny is better known for its garden attributes than as a wild flower. Its tight mats of rounded, fleshy leaves, illuminated by solitary, cup-like 25mm flowers, make it a popular ground-cover plant, concealing a multitude of garden sins. It sets little seed, but compensates by multiplying via vigorous creeping stems, anchoring itself to the soil by roots that emerge from buds at the leaf axils.

Historically, there was great faith in creeping Jenny's healing properties. Bruised leaves and an ointment made from the foliage were used to mend wounds. A carpet of the cultivar 'Aurea' lights up the garden with its glowing yellow foliage.

Greater bird's-foot trefoil [5]

Lotus pedunculatus (syn. *L. uliginosus*; Fabaceae or Leguminosae)

Perennial. Height Up to 1m. **Habitat** Damp, grassy habitats and marshes. **Distribution** All northern Europe as far south as southern Sweden. **Flowers** June–August. **Image size** Two thirds life-size.

The clusters of bright yellow pea flowers, typical of the bird's-foot trefoil species (see pp.91 and 100), are borne at the tips of tall, hollow stems. It is 'greater' than its relatives because the clusters may have up to 12 individual flowers, 10–18mm long, while the other species average only four to seven. Each leaf consists of three blue-green leaflets, with an extra pair of stipule-like leaflets at the leaf nodes. The original species name *uliginosus* – 'of swamps' – indicates its natural habitat.

Meadow buttercup [6]

Ranunculus acris (Ranunculaceae)

Perennial. Height Up to 1m. **Habitat** Damp, grassy habitats on chalky soil. **Distribution** All northern Europe. **Flowers** April–September. **In gardens** Moist, well-drained soil in full sun or partial shade. Borders and wild-flower meadows. **Image size** Two-thirds life-size.

A quintessential flower of summer meadows, children test their playmate's fondness for butter by looking for a yellow glow when held under the chin. This species is easily confused with the bulbous (p.88) and creeping (p.115) buttercups. All have leaves with two to seven deeply divided lobes but, unlike the other two species, the meadow buttercup's central lobe has no separate stalk. The flowers range from bright to paler shades of yellow and sometimes white. Like all buttercups, this species is poisonous; the leaf sap has a warning acrid taste.

The name buttercup came into general use in the 18th century. There is an Irish May Day tradition of rubbing the flowers on cows' udders to encourage milk production.

Smooth hawk's-beard [1]

Crepis capillaris (Asteraceae or Compositae)

Annual. Height 75cm. **Habitat** Grassland and wasteland. **Distribution** Britain and Ireland to Holland and Germany, naturalised in Denmark and Sweden. **Flowers** June–November. **Image size** One-third life-size.

Most hawk's-beard species look very similar, with the leaves being the easiest way of telling one from another. Those of smooth hawk's-beard are lanceolate and lobed; the basal leaves have stalks, but the upper ones clasp the stems. The flower heads are arranged in loose, branched clusters; each 10–15mm head – smallest of all the hawk's-beards – consists of golden strap-like rays with two rows of bracts beneath. The flowers are pollinated by bees and flies, and are followed by brown, ribbed fruits, which are carried on the wind by white hairs.

Bog asphodel [2]

Narthecium ossifragum (Liliaceae)

Rhizomatous perennial. Height Up to 45cm. **Habitat** Bogs, acid heaths and moors. **Distribution** Most of northern Europe except far north. **Flowers** July–September. **Image size** Life-size.

The bog asphodel is a delight, its creeping roots establishing it in drifts alongside other bog plants such as the tiny sundews (p.72) and marsh St John's wort. Its erect spikes of fragrant, golden-yellow starry flowers turn dark orange as they mature. Even then the show isn't over, as dramatic rich orange spiky fruits continue the display throughout autumn. The 15mm flowers have six pointed, ribbed petals and six stamens, which are woolly, with flame-coloured tips. There are a mass of strap-like leaves at the base of the plant and short stem leaves that clasp the flower stalks. Today populations have declined where its damp habitats are drained for cultivation. The Latin *ossifragum* means 'bone breaker'; the plant was unjustly blamed for causing brittle bones in sheep.

Hoary mustard [3]

Hirschfeldia incana (Brassicaceae or Cruciferae)

Annual or short-lived perennial. Height Up to 1.2m. **Habitat** Roadsides and wasteland. **Distribution** An introduction and alien from wool imports, naturalised in Britain, Belgium, Holland, France, Germany and Denmark. **Flowers** June–September. **Image size** One-quarter life-size.

A native of southern Europe, hoary mustard was cultivated farther north over 200 years ago and was recorded in the wild in the early 19th century. It has a rosette of pinnate, or pinnately lobed, basal leaves; the upper leaves are narrow and stalkless. Both the lower leaves and stem are densely covered with short, stiff white hairs. The pale yellow flowers give way to upright fruits, flattened against the stem, that have a short beak. There are three to six reddish-brown seeds on each side of the pod.

Fragrant agrimony [4]

Agrimonia procera (Rosaceae)

Perennial. Height Up to 1m. **Habitat** Grassland, fields and hedgerows, mostly on acid soils. **Distribution** Most of northern Europe except Faeroe Is, Iceland and northern Finland. **Flowers** June–August. **Image size** Life-size.

This is similar to agrimony (p.103) in general appearance, but is usually a more robust, well-branched plant with slightly larger flowers. The leaves are covered in small, shiny glands, which give off a fragrance when the foliage is bruised. The fruits are also larger, with the lowest spines bent sharply downwards – a distinguishing feature. The hooked bristles hang on to fur or clothing, aiding dispersal. This species may have arisen by a spontaneous doubling of the number of the common plant's chromosomes. It is found in very similar places, but is more scattered.

Mountain pansy [5]

Viola lutea (Violaceae)

Creeping perennial. Height Up to 20cm. **Habitat** Grassy or rocky upland, on alkaline or acid soil. **Distribution** Britain, Belgium, France, Holland and Germany. **Flowers** June–August. **In gardens** Well-drained soil, full sun or partial shade. Borders, rock gardens **Image size** Life-size.

Many upland limestone meadows are enriched by the mountain pansy's flat-faced flowers, usually alongside a mix of other species. It has creeping underground stems, which send up flower shoots at intervals. The flowers are larger (up to 30mm across) and more striking than wild pansy's (p.182), and there are several colour variants. These are not normally found together in the same place; in some areas, the yellow form is commonest, while in the mountains of Scotland, for example, most are the purple form. There have been attempts to bring the plant into cultivation since at least the 16th century. Several varieties are now available, but they are difficult to keep going.

Common meadow rue [6]

Thalictrum flavum (Ranunculaceae)

Perennial. Height Up to 1.2m. **Habitat** Damp habitats, meadows, fens, streamsides. **Distribution** Most of northern Europe except Iceland and Faroe Is. **Flowers** June-August. **In gardens** Any soil in full sun or partial shade. Herbabceous borders; wild and woodland gardens. **Image size** Life-size.

The fragile and diaphanous thalictrums have an enduring charm. Although related to buttercups, they do not look at all like buttercups. Common meadow rue has tall, wiry, branched stems, with fluffy clusters of clear yellow flowers at their tips. These consist simply of prominent stamens – the petals are very small and the sepals absent altogether. Despite lacking nectar, they attract bees and flies for their pollen; however, they are also wind-pollinated. The sparse leaves consist of three wedge-shaped, lobed leaflets, which are dark green.

Dark mullein [1]
Verbascum nigrum (Scrophulariaceae)

Biennial or short-lived semi-evergreen perennial.
Height Up to 1.2m. **Habitat** Wasteland and grassland on chalky soil. **Distribution** Northern Europe as far north as southern Sweden, except Ireland. **Flowers** July–October. **In gardens** Well-drained soil in full sun. Perennial borders. **Image size** Two–thirds life-size.

This perennial cousin of great mullein (see below) is altogether more compact. Its slim spikes are tightly packed with bright yellow saucer-like flowers up to 25mm across; the stems of the stamens have violet hairs. The oval, toothed leaves are hairless and dark green on the surface, pale and slightly woolly beneath. They have long stalks, which are shorter on the upper leaves.

Great mullein, Aaron's rod [2]
Verbascum thapsus (Scrophulariaceae)

Biennial. Height 1.2–2m. **Habitat** Grassland, hedgerows and roadsides. **Distribution** All northern Europe except far north. **Flowers** June–August. **In gardens** Well-drained soil in full sun. Wild-flower borders. Subject to mildew. **Image size** Two-thirds life-size.

In its second year, great mullein forms towering spikes of sulphur-yellow flowers – all the more striking for its silver-grey stems and woolly leaves. The name Aaron's rod refers to the Old Testament story of the staff that sprouted with blooms. The flowers, with five rounded petals, can reach up to 35mm across and attract bees and flies. They open randomly so that, at any one time, a single stem carries flowers at varying stages. The basal leaves vary from elliptical to oblong, and may have toothed edges. Farther up, the leaves, which are stemless, significantly smaller and more upright, hug the stem, which allows them to funnel rainwater to the base of the plant. The thick coating of hairs on both leaves and stem help the plant to retain moisture and deters browsing animals, whose mouths are irritated by the woolly texture.

Sea plantain [3]
Plantago maritima (Plantaginaceae)

Perennial. Height Up to 30cm. **Habitat** Coastal and damp rocky, mountainous habitats. **Distribution** All northern Europe. **Flowers** June–August. **Image size** Two-thirds life-size.

As in most other plantains, the yellow colour of the sea plantain comes from the stamens rather than the flowers themselves, which are brownish. Each flower head consists of a column of tiny flowers on a tall, smooth stem above a straggly rosette of narrow, fleshy leaves. It is one of the few plants that can tolerate a high level of salt in water and it thrives in salt marshes. Sheep love the leaves, and in Wales it was once cultivated as fodder.

Field pansy [4]
Viola arvensis (Violaceae)

Annual. Height 10–40cm. **Habitat** Wasteland and arable land on neutral or chalky soil. **Distribution** All northern Europe except far north. **Flowers** April–October. **Image size** Two-thirds life-size.

Often unfairly dubbed a weed simply because it inhabits arable land, the field pansy is a pretty, delicate plant. It is variable, usually producing pale creamy-yellow 4–8mm flowers, but the petals are sometimes painted with purple blotches. Like other wild pansies, the sepals are longer than the petals. The elongated leaves have blunt teeth or scalloped edges. When heartsease (p.182) grows nearby, the two are prone to bouts of promiscuity, producing a range of purple, cream and/or yellow colour variants.

Pineappleweed [5]
Matricaria discoidea (syns. *Chamomilla suaveolens* and *Matricaria matricarioides*; Asteraceae)

Annual. Height Up to 35cm. **Habitat** Wasteland and paths. **Distribution** Introduction. Naturalised in most of northern Europe except Faroe Islands and Iceland. **Flowers** June–July. **Image size** Two-thirds life-size.

Without any semblance of petals, the globular dirty yellow flower heads of the pineappleweed look rather naked. What the flowers lack in beauty, however, is compensated by the bright green feathery leaves, which have a pineapple scent and pepper the stocky, branched stems. It is native to north-east Asia, but was introduced to Europe from North America. It is tolerant of being crushed so has made itself at home along well-worn tracks and paths.

Wild turnip [6]
Brassica rapa (Brassicaceae or Cruciferae)

Annual or biennial. Height Up to 1m. **Habitat** River- and canal-banks, waste and arable land. **Distribution** Probably introduced; all northern Europe. **Flowers** May–August. **Image size** Half life-size.

This wild form of the cultivated vegetable has grass-green leaves and yellow flowers in heads that carry open flowers above the unopened buds. There are three distinct sub-species or varieties. Pineapple mayweed is slender-rooted, and has all the look of a native plant. The second, also slender-rooted, has red-brown seeds and is grown as an oil-seed plant and for fodder. The third is the classic swollen-rooted vegetable.

Mugwort [7]
Artemisia vulgaris (Asteraceae or Compositae)

Perennial. Height 80–150cm. **Habitat** River-banks, wasteland, roadsides and hedgerows. **Distribution** Most of northern Europe. **Flowers** May–September. **Image size** Half life-size.

Despite its common name and scruffy appearance, mugwort has been associated with magic for 1,000 years. The reddish, grooved stems form upright clumps. Each branch is smothered in clusters of flowers consisting of tiny florets enveloped in bracts. The deeply cut foliage, which is covered in a soft woolly down, is a dull green on the upper surface and silvery underneath.

Golden samphire [1]

Inula crithmoides (Asteraceae or Compositae)

Perennial. Height Up to 1m. **Habitat** Coastal habitats including salt marshes, shingle and cliffs. **Distribution** France, southern and western Britain and Ireland. **Flowers** July–September. **Image size** Two-thirds life-size.

Narrow, fleshy foliage is golden samphire's secret weapon for surviving the dry and salty conditions of its various coastal habitats. The yellow-green smooth leaves are either untoothed or have a three-toothed tip. Conspicuous clusters of luminous yellow, somewhat daisy-like flowers grow at the tip of branched stems, which form lax clumps. Golden samphire is not related to rock samphire (p.104) and, unlike that edible species, has no significant culinary uses.

Ploughman's spikenard [2]

Inula conyza (Asteraceae or Compositae)

Biennial or perennial. Height Up to 1.25m. **Habitat** Wood edges, grassland, rocky and disturbed areas, generally on chalky soils. **Distribution** Northern Europe as far north as England, Wales and Denmark; an introduction in Ireland. **Flowers** July–September. **Image size** Two-thirds life-size.

Spikenard was a salve for dressing wounds much prized by the Romans; it was extracted from the fragrant roots of an Indian plant of the genus *Nardostachys*. Ploughman's spikenard is unrelated, but it also has fragrant roots. These would have been noticed by a ploughman as the plough disturbed and broke them, and the plant acquired its name by association. At first sight of the basal leaves alone, you might mistake it for a foxglove, but the clusters of flower heads at the top of the tall, branching, purplish stems easily distinguish it. Each head is about 10mm long and consists mainly of disc florets, with an outer ring of very short ray florets – both types a dull yellow. The outer flower bracts are green, the inner ones purplish. When seeds are set, each has a parachute of reddish-white hairs.

Elecampane [3]

Inula helenium (Asteraceae or Compositae)

Perennial. Height Up to 2.5m. **Habitat** Wasteland, orchards and roadsides. **Distribution** All northern Europe. **Flowers** July–August. **In gardens** Moist, well-drained soil in full sun. Herb gardens and perennial borders. **Image size** Two-thirds life-size.

An ancient herb, elecampane was boiled, dried and infused to make a cure for medical complaints as well as a remedy for snake-bites and a pick-me-up for horses. An extract, inulin, has recently been used in conventional medicine to treat asthma. In Scandinavia, elecampane was woven into the centre of a posy of herbs to symbolise the sun and the head of the Norse god Odin. It is a tall, untidy plant with sunflower-like flower heads up to 80mm across on thick, downy stems. The large oval leaves have toothed edges and clasp the top of the flower stem. In gardens, keep plants in check by dividing and replanting them in autumn or cutting back to stop them self-seeding.

Hop trefoil [4]

Trifolium campestre (Fabaceae or Leguminosae)

Annual. Height 10–30cm. **Habitat** Grassland, roadsides and sand dunes. **Distribution** All northern Europe except Iceland. **Flowers** June–September. **Image size** Two–thirds life-size.

Each hop trefoil flower head consists of numerous tiny yellow flowers, each with a broad upper petal with a furrowed surface. When the flower heads dry they turn brown and papery, and remain attached to the stem; they look just like miniature hops. The flower heads appear at the tip of branched stems, which are covered in down and carry leaves consisting of three leaflets. When in flower, hop trefoil is easily mistaken for black medick (p.100), but there should be no confusion between the hop-like seed heads and the distinctive coiled black pods of the black medick.

Hawkweed oxtongue [5]

Picris hieracioides (Asteraceae or Compositae)

Biennial or perennial. Height Up to 1m. **Habitat** Grassland and bare ground on chalky soil. **Distribution** Northern Europe except Faroe Is, Iceland and Norway. **Flowers** July–October. **Image size** Two-thirds life-size.

Hawkweed oxtongue is very closely related to the bristly oxtongue (p.96), which causes confusion since both plants grow in similar habitats. Both have bright yellow dandelion-like flower heads, which are followed by seeds carried on the wind by hairy parachutes. However, the hawkweed oxtongue flower heads have sepal-like bracts that are narrower and covered in black hairs. Both species have arrow-shaped leaves covered in a layer of bristly hairs, but those of hawkweed oxtongue have toothed rather than wavy edges and a pinkish-red midrib. The genus name *Picris* comes from the Greek *pikros*, and refers to the bitter taste of the milky sap that exudes from cut stems.

Reflexed stonecrop [6]

Sedum rupestre (syn. *Sedum reflexum*; Crassulaceae)

Evergreen perennial. Height Up to 35cm. **Habitat** Stony banks, walls and rocks. **Distribution** All northern Europe except Iceland and far north. **Flowers** June–August. **In gardens** Well-drained soil in full sun. Rock gardens. **Image size** Two-thirds life-size.

A native of central and western Europe, reflexed stonecrop has escaped from gardens elsewhere and naturalised with ease, either by propagating itself from small pieces of plant or producing masses of seed. Upright succulent stems grow from spreading mats of grey-green fleshy foliage and carry clusters of sulphur-yellow, star-like flowers. This species is very similar to the rock stonecrop (p.107) except for the foliage, which has a more relaxed habit and falls to the ground as soon as it dies. Reflexed stonecrop is so called because the leaves at the base of the flower stems curve backwards. The edible leaves were eaten in salads.

Yellow loosestrife [1]
Lysimachia vulgaris (Primulaceae)

Perennial. Height Up to 1.2m. **Habitat** Marshes, ditches and lake- and river-banks. **Distribution** Most of northern Europe except Iceland and Faeroe Is. **Flowers** June–August. **In gardens** Moist, well-drained soil in full sun or partial shade. Damp herbaceous borders and bog gardens. **Image size** Two-thirds life-size.

When the cupped starry flowers open in early summer, it is clear why yellow loosestrife made such an easy transformation from wild flower to garden plant. It makes attractive clumps, with pyramids of 20mm blooms, each with five petals, on side-shoots from the branched stems. The stems are covered in a layer of pale grey woolly hairs. The sepals have orange edges, while the oval leaves are dotted with orange or black glands. The flowers have no scent and produce no nectar to attract insects. Despite this, they are pollinated by a bee species, *Macropis labiata*, which visits only this plant.

The name *Lysimachia* may come from Greek words meaning to 'loosen strife', from its use to prevent insects irritating oxen. It may also be named after Lysimachus, King of Sicily, who supposedly discovered its medicinal properties.

Celery-leaved crowfoot or buttercup
[2] *Ranunculus sceleratus* (Ranunculaceae)

Annual. Height Up to 60cm. **Habitat** Marshes, ponds, streamsides and damp meadows. **Distribution** All northern Europe except far north. **Flowers** May–September. **Image size** Two-thirds life-size.

S*celeratus* means 'vicious' and celery-leaved crowfoot can fatally poison livestock; the sap blisters human skin. The shiny, celery-like lobed leaves are dark green. As in the meadow buttercup (p.107), the middle leaf lobe has no separate stalk. The pale yellow flowers are smaller than in other *Ranunculus* species – 5–10mm across. They appear at the branch tips of stout grooved stems and are followed by elongated fruits.

Creeping buttercup [3]
Ranunculus repens (Ranunculaceae)

Creeping perennial. Height Up to 60cm. **Habitat** Grassy habitats and roadsides. **Distribution** All northern Europe. **Flowers** May–September. **In gardens** A pernicious weed. **Image size** Two-thirds life-size.

It may have the familiar yellow cup-shaped flowers and typical lobed leaves, but the creeping buttercup differs from other buttercup species (pp.88 & 107) by having tenacious creeping roots. It is the bane of gardeners' lives once it gets a grip in a lawn, lying low enough to avoid the lawnmower's blade. It's tough, too, happily putting up with being trampled underfoot. If you catch it early enough, pull the plant out by hand.

Perennial sow thistle [4]
Sonchus arvensis (Asteraceae or Compositae)

Perennial. Height 80–150cm. **Habitat** Waste and arable land, dunes and river-banks. **Distribution** All northern Europe except far north. **Flowers** July–October. **Image size** Two-thirds life-size.

The handsome flower heads of perennial sow thistle grow up to 50mm across and consist of numerous strap-shaped florets. They grow on upright, hollow stems filled with a milky sap and are followed by fluffy dandelion-like seed heads. Long leaves, arranged alternately up the stems, have spiny edges and rounded lobes at the base that clasp the stem. As with all sow thistles, they are edible and used to be eaten in salads or fed to livestock.

Tansy [5]
Tanacetum vulgare (Asteraceae or Compositae)

Perennial. Height Up to 1.2m. **Habitat** Grassland and wasteland. **Distribution** Most of northern Europe. **Flowers** July–September. **In gardens** Well-drained soil in full sun. Herb gardens, borders and containers. **Image size** Two-thirds life-size.

At dusk, tansy's branched clusters of yellow button-like flowers look like luminous torches on top of their tall grooved stems. The 10mm flower heads consist solely of disc florets. The ferny leaves, with up to 12 toothed leaflets, are aromatic; they were used to spice up omelettes and cakes. (Tansy cake was traditional at Easter.) It was also used to repel blowflies on meat and corpses. Tansy forms neat clumps, but it is invasive, so in gardens it is best confined to a container.

Common fleabane [6]
Pulicaria dysenterica (Asteraceae or Compositae)

Perennial. Height Up to 1m. **Habitat** Marshes, riversides and wet meadows. **Distribution** Most of northern Europe except far north. **Flowers** August–September. **Image size** Two-thirds life-size.

Common fleabane makes up for its lack of beauty with formidable insecticidal powers. As the common and botanical names suggest – *Pulicaria* comes from the Latin *pulex* (flea) – it was used to keep fleas at bay. The pale yellow daisy-like flowers are up to 30mm across, and open at the tips of upright, branched stems. Spear-shaped woolly leaves grow alternately all the way up the woolly stem.

Hoary ragwort [7]
Senecio erucifolius (Asteraceae or Compositae)

Biennial or perennial. Height Up to 1.2m. **Habitat** Grassland and roadsides. **Distribution** All northern Europe except Norway and Finland. **Flowers** July–September. **Image size** Two-thirds life-size.

Hoary ragwort is very similar to the better known common ragwort (p.100), with the same yellow daisy-like flowers on upright branched stems. Hoary ragwort has fewer flower heads in each cluster, and the leaves are less lobed and have woolly undersides. Like all ragwort species, it is fatally poisonous to livestock.

Goldenrod [1]

Solidago virgaurea (Asteraceae or Compositae)

Perennial. Height Up to 70cm. **Habitat** Grassland, open woodland and cliffs. **Distribution** All northern Europe. **Flowers** July–September. **In gardens** Well-drained soil in full sun. Rock gardens and borders. **Image size** Two-thirds life-size.

This is the only *Solidago* species out of 80 that is native to northern Europe; the rest come from North America (p.119) and other parts of Europe and Asia. The rather ragged bright yellow flower heads, 15–18mm across, consist of two untidy layers of disc and ray florets, and appear in large branched clusters. Toothed leaves grow in spirals up the stems. *Solidago* means 'to make whole' and, used as an ointment or taken internally, the plant had a great reputation for healing wounds in the late Middle Ages. As a result, it was highly sought-after until it was found growing wild in London. After that, the herbalist John Gerard wryly noted, 'No man will give halfe a crowne for an hundred weight of it.'

Nipplewort [2]

Lapsana communis (Asteraceae or Compositae)

Annual. Height Up to 1m. **Habitat** Wasteland, roadsides and hedgerows. **Distribution** All northern Europe. **Flowers** June–October. **In gardens** A troublesome weed. **Image size** Two-thirds life-size.

Several centuries ago someone with an entrepreneurial spirit thought they saw a resemblance between the closed flower buds of this plant and nipples. It was christened nipplewort and used to heal sore nipples. Today its only real significance is as a pernicious garden weed. Yet, like its relative the dandelion, the nipplewort's foliage is edible and makes a tasty green salad. It is a sparse plant with wiry branched stems and insignificant yellow daisy-like flowers of strap-like florets, which only bother to open in bright sunshine. The oval leaves have toothed edges and a wing at their base.

Eastern rocket [3]

Sisymbrium orientale (Brassicaceae or Cruciferae)

Annual. Height Up to 1m. **Habitat** Wasteland. **Distribution** Introduction. Naturalised in most of northern Europe. **Flowers** June–August. **Image size** Two-thirds life-size.

Eastern rocket was introduced from Africa and western Asia. It has the typically delicate four-petalled flowers of the cabbage family. The petals are obviously larger than the sepals and they appear at the same time as the slender, hairy seedpods of earlier flowers. These pods can reach up to 5cm long. The plant's foliage distinguishes it from several similar species in the same family. It is much like tall rocket (*Sisymbrium altissimum*), but the leaves are more robust and the flowers packed less densely. Eastern rocket's stems are covered in soft hairs, as are the leaves, all but the uppermost of which are lobed.

Groundsel [4]

Senecio vulgaris (Asteraceae or Compositae)

Annual. Height Up to 30cm. **Habitat** Wasteland and cultivated soil. **Distribution** All northern Europe. **Flowers** Year-round. **In gardens** A weed. **Image size** Two-thirds life-size.

Groundsel is closely related to the dreaded ragworts (pp.100 & 115). It is not viciously poisonous like those plants, but is just as much a problem simply because it multiplies so rapidly and vigorously. The seeds are carried on the wind by parachutes of silky hairs. On touching down on any available clear ground, they make a rapid advance – hence the name groundsel, which comes from the Anglo-Saxon *grondeswyle*, meaning 'ground swallower'. The yellow flower heads, usually consisting of tiny florets, are sheathed in black-tipped bracts. Occasionally, a flower emerges with a few more prominent outer florets to look more like a yellow daisy. The leaves, which are arranged spirally up the purple stems,

have irregular toothed lobes. Today groundsel is used as food for caged birds. If it threatens to take over the garden, pull it out by hand and compost it.

Autumn hawkbit [5]

Leontodon autumnalis (Asteraceae or Compositae)

Perennial. Height Up to 60cm. **Habitat** Grassland, roadsides and upland screes. **Distribution** All northern Europe except Faeroe Is and Iceland. **Flowers** June–October. **Image size** Half life-size.

Autumn hawkbit often produces several rosettes of leaves from a single much-branched root system. The stems arising from them are hairy and usually branched, with scale-like bracts just below each flower head. These heads consist of golden-yellow ray florets, the outer ones streaked reddish on the underside. In mountain areas forms of autumn hawkbit are found where the bracts of the flower heads are covered in black, woolly hairs. The leaves are oblong-lanceolate in shape but vary from almost toothless to deeply divided and pinnately lobed. The seeds are brown, with a parachute consisting of a single row of hairs.

Canadian goldenrod [1]

Solidago canadensis (Asteraceae or Compositae)

Perennial. Height 1–2.5m. **Habitat** Grassland, wasteland and banks. **Distribution** Introduction. Widely naturalised in northern Europe. **Flowers** August–October. **In gardens** Well-drained soil in full sun. Wild-flower gardens. **Image size** Life-size.

Canadian goldenrod was introduced to northern Europe from North America in the mid-17th century. Later, the majestic upright stems of bright yellow flower heads became a familiar sight in Victorian cottage gardens. However, its prolific ability to self-seed ensured that it soon escaped into the wild, where it tends to colonise railway embankments and sidings, and other wasteland. The flowers, like tiny golden-yellow daisies, are attractively arranged along near horizontal branches making great golden pyramids. The long, lanceolate leaves have pointed tips and slightly toothed edges.

Prickly lettuce [2]

Lactuca serriola (Asteraceae or Compositae)

Annual or biennial. Height Up to 1.8m. **Habitat** Wasteland and sand dunes. **Distribution** Most of northern Europe except Ireland, Scotland and the far north. **Flowers** July–September. **In gardens** A weed. **Image size** One-third life-size.

Prickly lettuce bears no resemblance to our salad vegetable, although it belongs to the same genus. The long, lobed foliage is armed with unappetising and menacing prickles on both the margins and the midrib beneath and clasps prickly upright stems. It tends to align north–south, so is sometimes called the 'compass plant'. Prickly lettuce can be mistaken for the great lettuce (*Lactuca virosa*), but has more prickles at the base of the leaves and the fruits are grey rather than dark maroon. Birds love the seeds, so it is worth leaving a few plants in the garden despite their untidy appearance.

Wall lettuce [3]

Mycelis muralis (Asteraceae or Compositae)

Perennial. Height Up to 1m. **Habitat** Shady places on rocks and walls, or in woodland, usually on chalky soil. **Distribution** All northern Europe except Iceland and Faeroe Is. Naturalised in Ireland. **Flowers** July–September. **Image size** One-third life-size.

At first sight, wall lettuce is an untidy-looking plant. The lower leaves are long-stalked and pinnately lobed, with the side lobes diamond-shaped and the end lobe triangular; they are edible and can be used in salads. The flower heads are borne on the end of short branches, in open panicles. Each head is small, consisting of three to five pale yellow ray florets, and is surrounded by an outer ring of blunt, spreading bracts, which are usually reddish. The small seeds are blackish, with a short pale beak. Plants growing on relatively infertile soil are healthiest.

Sharp-leaved fluellen [4]

Kickxia elatine (syn. *Linaria elatine*; Scrophulariaceae)

Creeping annual. Height Up to 20cm; stems up to 50cm long. **Habitat** Arable or disturbed land on chalky or sandy soil. **Distribution** Throughout most of northern Europe except Iceland, Faeroe Is and Norway. A long-established introduction in Britain. **Flowers** July until frosts. **Image size** Twice life-size.

The two European native species of fluellen are closely related to toadflax (p.107), and look very like it. There are, however, two main differences: fluellen leaves tend to be broader than those of toadflax; and the small snapdragon-like flowers are solitary rather than arranged in spikes. The British common name comes from the Welsh *ilysiau Llewelyn* – 'Llewelyn's flower' – first used for a species of veronica. The genus name commemorates Jean Kickx, a 19th-century apothecary of Brussels. Sharp-leaved fluellen is a scrambling plant, often with many branches on a stem. It has yellow two-lipped flowers with a violet upper lip

and a straight spur. They grow on virtually hairless sprawling stems rising from leaf nodes. The leaves are hairy, arrow-shaped and sharply pointed, or almost triangular.

Yellow mountain saxifrage [5]

Saxifraga aizoides (Saxifragaceae)

Perennial. Height Up to 25cm. **Habitat** By water and on wet, stony ground on mountains. **Distribution** All northern Europe except Faeroe Is, but especially Arctic and sub-Arctic. **Flowers** June–September. **In gardens** Well-drained but moist soil in sun. Rock gardens and alpine troughs. **Image size** Two-thirds life-size.

Yellow mountain saxifrage is most at home by small rocky mountain streams, where the stems clamber over stones, often forming dense mats. The leaves are fleshy, stalkless and narrowly oblong to prevent water loss during long periods of freezing conditions. Upright flower stems bear loose heads of up to ten flowers, which have spreading, triangular sepals and five narrow yellow petals, often red-spotted. Yellow saxifrage also grows happily at sea level in northern parts, where the climate at sea level can be as harsh as in the mountains.

Round-leaved fluellen [6]

Kickxia spuria (syn. *Linaria spuria*; Scrophulariaceae)

Creeping annual. Height Up to 50cm. **Habitat** Arable land. **Distribution** Belgium, southern Britain, France, Germany and Holland. **Flowers** July–October. **Image size** Two-thirds life-size.

Round-leaved fluellen is very similar to the sharp-leaved species (see left), but its foliage is oval or sometimes heart-shaped. The yellow flowers have a deep purple rather than violet upper lip, a curved spur and a hairy stalk. The two species sometimes hybridise, producing plants with pointed leaves and hairy flower stalks. Like many annuals, fluellens often self-pollinate and set seed without opening at all.

Charlock [1]

Sinapis arvensis (Brassicaceae or Cruciferae)

Annual. Height Up to 1m. **Habitat** A pernicious weed of arable land; also on waste ground, rubbish tips and roadsides. **Distribution** Throughout northern Europe except far north. **Flowers** May–July. **Image size** One-quarter life-size.

Charlock is a very successful weed. Before the widespread use of selective weedkillers, it could take over a field of spring-sown cereal. Its seed can lie dormant quite happily for at least ten years, and there are reports of it germinating in pasture that has been unploughed for as long as fifty years. Once disturbed, charlock germinates rapidly and proceeds with its hostile takeover. It is a roughly hairy annual, which may be branched or unbranched. The pale yellow flowers have four sepals spreading horizontally below the petals. Its seed pods are cylindrical and have a long beak; the red-brown seeds are in two rows, on each side of a dividing septum (membrane or wall). The lower leaves are large and lyre-shaped, the upper ones smaller and lanceolate.

Yellow bartsia [2]

Parentucellia viscosa (Scrophulariaceae)

Annual. Height 30–50cm. **Habitat** Damp grassland, often near sea. **Distribution** Britain, Ireland and France; naturalised in Belgium, Holland and Denmark. **Flowers** June–September. **Image size** Half life-size.

The 18th-century botanist Carl Linnaeus named bartsias to commemorate the death of his friend Dr Johann Bartsch, but only a few of the closely related species are now classified in his original genus *Bartsia*. Yellow bartsia is a stocky, upright plant with stiff stems. The lanceolate leaves are toothed and arranged in pairs. Pale yellow flowers emerge at the leaf axils. They have two lips – the upper hooded and the lower with three lobes. The whole plant is stickily hairy all over. In Britain, it mostly thrives in the south-west; in Devon it is called tweeny-legs.

Silverweed [3]

Potentilla anserina (Rosaceae)

Creeping perennial. Height 5–25cm. **Habitat** Grassland, sand dunes, wasteland, roadsides and tracks. **Distribution** All northern Europe. **Flowers** May–August. **Image size** Half life-size.

Silverweed is one of the most common of nearly 500 *Potentilla* species in northern Europe. It has long trailing (and self-rooting) stems and pinnate leaves, each consisting of up to 25 leaflets with toothed edges. They are covered in silvery hairs – hence the common name (*silberkraut* in German, and *zilverkruid* in Dutch) – and can form a dense mat. The yellow flowers, 20mm across, have five rounded petals.

Soft foliage and flexible stems let silverweed tolerate trampling underfoot but don't protect it from livestock and geese, who love to eat it; *anser* is Latin for 'goose'. *Potentilla* means 'powerful', and refers to the many and varied medicinal properties attributed to this genus. More mundanely, the hairy leaves were stuffed into travellers' boots to relieve weary feet. It has culinary uses too, and in times of hardship silverweed roots, which taste of parsnip, were a substitute for potatoes.

Navelwort [4]

Umbilicus rupestris (Crassulaceae)

Perennial. Height Up to 40cm. **Habitat** Rocks, walls, cliffs and rooftops. **Distribution** Britain and Ireland (except far north), France. **Flowers** June–August. **Image size** Two-thirds life-size.

Each fleshy disc-like leaf of the navelwort, with its central depression, bears a close resemblance to a belly button. The Romans associated it with Venus's navel and with lovemaking. The individual leaves are borne on long, succulent stems. The bell-shaped 8–10mm flowers are greenish-white and appear in spikes, each on a short drooping stem. The succulent nature of navelwort allows it to survive in less than hospitable rocky habitats. (*Rupestris* means 'growing on rocks'.) It is also known as pennywort, referring to the rounded foliage.

Weld, dyer's rocket [5]

Reseda luteola (Resedaceae)

Biennial. Height Up to 1.3m. **Habitat** Stony ground, grassy and sandy habitats and wasteland, usually on chalky soils. **Distribution** Patchily throughout northern Europe except far north. **Flowers** June–September. **Image size** Two-thirds life-size.

Weld is also known as dyer's rocket because it was the source of a bright yellow fabric dye used as long ago as the Stone Age. The yellowish-green flowers appear in long, slender spikes which, from afar, look like magnificent tapered candles. They are heliotropic – meaning that they turn to face the sun, tracking it across the sky from east to west. The lanceolate leaves have wavy margins and are arranged in a rosette at the base but also grow alternately up the length of the stem. The genus name *Reseda* comes from the Latin *resedo* (to heal); *luteola* means 'yellowish', referring to the flower colour.

Buck's-horn plantain [6]

Plantago coronopus (Plantaginaceae)

Biennial or perennial; occasionally annual. Height 10–25cm. **Habitat** Short grassland and coastal sandy and shingly habitats. **Distribution** All northern Europe except the far north. **Flowers** May–July. **Image size** Two-thirds life-size.

The 40mm spikes of yellowish-brown flowers, with their prominent yellow stamens, set it apart from its relatives hoary and ribwort plantains (p.43); these both have shorter and more rounded flower heads. The name buck's-horn refers to the elongated, divided leaves which, with their prominent pointed lobes, resemble stags' horns. In France, the mucilage that exudes from the plant's wet seeds was used to stiffen fabric.

Monkey flower [1]
Mimulus guttatus (Scrophulariaceae)

Perennial. Height Up to 75cm. **Habitat** Marshy habitats; river and lake margins. **Distribution** Introduction in most of northern Europe. **Flowers** July–September. **In gardens** Fertile, moist soil in full sun. Bog gardens and borders. **Image size** Two-thirds life-size.

The monkey flower's flamboyant good looks hint at exotic origins. It was first brought to Europe from North America in the early 19th century as a garden plant. But it spreads vigorously on creeping roots, so it took no time to escape into the wild. Each flower has an upper and lower lip with two bulges on the lower lip covered in red spots, providing a landing guide for pollinating insects. The flowers grow at the top of sturdy upright stems, from the leaf nodes. Their 'facial features' inspired both its common and Latin names – *Mimulus* meaning 'little actor' and *guttatus* meaning 'spotted'. The oval, coarsely veined leaves with toothed edges are arranged in pairs.

Ribbed melilot [2]
Melilotus officinalis (Fabaceae or Leguminosae)

Biennial or perennial. Height Up to 1.5m. **Habitat** Grassland and wasteland. **Distribution** Most of northern Europe but naturalised in Britain, Ireland and Scandinavia. **Flowers** July–September. **Image size** Two-thirds life-size.

Only slight differences distinguish one melilot species from another. The yellow flowers of the ribbed melilot are typical of the pea family but, in this case, the lower petal (the keel) is shorter than the rest. They are arranged in loose spikes on spindly stems, looking rather like miniature lupins. They brim with nectar, which attracts pollinating hoverflies and bees. The leaves consist of three leaflets, rather like clover, with sharp, toothed edges. The oval seedpods are ribbed and hairless – unlike tall melilot (p.99), which has hairy fruits. All the species give off a delicious scent of newly mown hay when dried.

Dragon's-teeth [3]
Tetragonolobus maritimus (Fabaceae or Leguminosae)

Prostrate perennial. Height Up to 30cm. **Habitat** Grassland (often damp) on chalky soil. **Distribution** Naturalised in southern England, France, Germany, Denmark and southern Sweden. **Flowers** May–September. **Image size** Two-thirds life-size.

Dragon's-teeth is related to common bird's-foot trefoil (p.91), and both have yellow pea-like flowers. The main difference between the two lies in the solitary flowers of dragon's-teeth, which are plain yellow and have a leaf-like bract below each bloom. Bird's-foot trefoil produces clusters of yellow flowers streaked with red. Dragon's-teeth has prostrate stems and trifoliate leaves, which may or may not be covered in soft downy hairs. The seedpods are distinctive; they can be 3–6cm long and are angled, with wings on each angle.

Fennel [4]
Foeniculum vulgare (Apiaceae or Umbelliferae)

Perennial. Height 1.5–2.5m. **Habitat** Waste and cultivated land and roadsides. **Distribution** Possible introduction to Britain and France. Naturalised in Belgium, Germany and Holland. **Flowers** July–October. **In gardens** Moist, well-drained soil in full sun. Herb gardens, borders and containers. **Image size** Two-thirds life-size.

Fennel bulbs, leaves and seeds, with their unmistakable aniseed aroma and flavour, have been widely used in cooking since Roman times. It is also now much loved as an ornamental garden plant, thanks to its diaphanous feathery foliage. The tiny flowers – held in flat umbels – are also pretty enough in a subtle shade of yellow that complements the stiff blue-grey stems. The whole plant of the wild species looks like a puff of blue-grey smoke and there are a couple of stunning ornamental cultivars: 'Giant Bronze' has copper-coloured foliage, which ages to a dark bronze, and 'Purpureum' has bronze juvenile foliage, which turns blue-grey as it matures.

Welsh poppy [5]
Meconopsis cambrica (Papaveraceae)

Perennial. Height 40–60cm. **Habitat** Shady habitats: woodland, hedgerows and rocky places. **Distribution** Southern and western Britain and Ireland and western France. Naturalised in some other areas. **Flowers** June–August. **In gardens** Moist, well-drained soil in sheltered partial shade. Wild-flower gardens and herbaceous borders. **Image size** Two-thirds life-size.

The papery-thin petals of the Welsh poppy are a pure, clear yellow; at the centre of each solitary cup-shaped flower – up to 80mm across – sits a crown of yellow stamens. The wiry, woolly flower stems emerge from tufts of sparsely hairy pinnate leaves with deeply lobed leaflets. This perennial species is very similar to the annual red poppies (p.160) and originally belonged to the same genus, *Papaver*. Only when it was noted that the seeds are released from the seed capsule through slits rather than from perforations at the top of a 'pepper-pot' capsule was it reclassified as *Meconopsis* – from Greek words meaning 'poppy-like'. The Welsh poppy is well known in gardens.

Wild radish [6]
Raphanus raphanistrum (Brassicaceae or Cruciferae)

Annual. Height Up to 75cm. **Habitat** Waste and cultivated ground. **Distribution** All northern Europe except far north; possibly introduced. **Flowers** May–September. **Image size** Two-thirds life-size.

Although it is related to the garden radish (*Raphanus sativus*), this bane of all farmers doesn't look like it or have any culinary value. The only clue is the unusual 'waisted' seedpods, up to 5cm long, each of which holds up to eight seeds. As a seed ripens, its segment of pod breaks off and falls to the ground. Wild radish makes untidy but sturdy clumps of bristly foliage. The wiry flower stems have a pinkish tinge and the cross-shaped flowers can be yellow, lilac or white. Whatever the flower colour, the petals have a fine tracery of purplish veins.

Lesser bladderwort [1]
Utricularia minor (Lentibulariaceae)

Aquatic perennial. Height Stems up to 25cm above surface. **Habitat** Still pools and ditches. **Distribution** All northern Europe except Faeroe Is. **Flowers** June–September. **Image size** One-and-one-third life-size.

At first sight, this is merely a smaller version of greater bladderwort, but it has underwater stems of two distinct types. One has finely dissected green leaves, which bear only a few bladders. The other lacks green chlorophyll and may be buried in mud at the bottom of the pond; this type of stem has bladders on much-reduced leaves. The pale greenish-yellow flowers are smaller than those of greater bladderwort, with a spur that is conspicuously short and blunt.

Sticky groundsel [2]
Senecio viscosus (Asteraceae or Compositae)

Annual. Height Up to 60cm. **Habitat** Walls, wasteland, roadsides and railway tracks. **Distribution** Britain, Ireland, Belgium, Holland, France, Germany. An introduction. **Flowers** July–September. **Image size** Half life-size.

This wayside annual has spreading branches, is extremely sticky – so that in dusty conditions dirt and debris attach to it – and has an unpleasant smell. The dark green leaves are pinnately lobed, with the lobes themselves divided; the lower leaves are short-stalked, the upper stalkless. The pale yellow flower heads are borne in a large, rounded inflorescence. Each head has central disc florets and up to 15 short ray florets around the edge; these roll back shortly after opening. The seeds have long white hairs – wind and traffic has spread the plant along roads and railways.

Greater bladderwort [3]
Utricularia vulgaris (Lentibulariaceae)

Aquatic perennial. Height Stems up to 45cm above surface. **Habitat** Still or slow-moving water. **Distribution**
All northern Europe except Iceland. **Flowers** July–August. **Image size** One-quarter life-size.

Like the sundews (p.72) and butterwort (p.189), bladderworts have evolved a unique method of obtaining vital food. They are free-floating plants with much-divided submerged leaves that bear small bladders. These are full of air and have bristles at one end. When a small creature such as a water flea, brushes against the bristles, a small trapdoor opens and water rushes into the bladder, pulling in the hapless flea. The bladders absorb the useful products as the insect decomposes.

The only part of the plant that appears above water is the spike of yellow flowers. These are two-lipped, with a spur on the corolla. Bladderworts have solved the problem of surviving winter by forming a detachable bud called a turion, which sinks to the bottom and lays dormant until the water warms up in spring. At this point, it surfaces and begins growing again.

Yellow water lily [4]
Nuphar lutea (Nymphaeaceae)

Aquatic perennial. Height Stems 10cm above water surface. **Habitat** Slow-moving rivers, lakes and ponds. **Distribution** All northern Europe except Faeroe Is and Iceland. **Flowers** June–September. **In gardens** Water up to 2m deep, in full sun. Large ponds and lakes. **Image size** One-quarter life-size.

Rather than floating on the water's surface, like other water lilies, the flowers of this species are held aloft on stout upright stems. The flowers themselves look more like 12–40mm aquatic buttercups rather than water lilies. They give off an alcoholic scent, which attracts pollinating flies. The underwater circular leaves are thin and translucent, forming a crimped cup. At the surface they are flat, reaching up to 40cm across, and are thick and leathery. In France the yellow water lily had a reputation for ruining one's sex drive. Beware: it can engulf the whole surface of a domestic pool.

Wormwood [5]
Artemisia absinthium (Asteraceae or Compositae)

Perennial. Height Up to 1m. **Habitat** Wasteland, roadsides and coastal habitats. **Distribution** Most of northern Europe. **Flowers** July–August. **In gardens** Well-drained soil in full sun. Rock gardens, herb gardens and borders. **Image size** Life-size.

The multi-talented wormwood was believed to keep goblins and evil spirits at bay. Most notoriously, it was the flavouring of the French spirit *absinthe*. Yet it still manages to look good enough to plant in the garden. Its most attractive feature is the mounds of finely cut, feathery foliage; it is a striking silvery-green colour and has a silky texture that invites you to run your fingers through it to sample its aroma. The flowers are unexciting dirty yellow buttons, tightly enclosed in green bracts, and appear late in the season. The British common name comes from the German *wermut*, which is also the origin of the name vermouth.

Fringed water lily [6]
Nymphoides peltata (Menyanthaceae)

Aquatic perennial. Height Stems up to 10cm above water surface. **Habitat** Shallow, slow-moving rivers and ponds. **Distribution** Britain, Belgium, France, Holland and Germany. Naturalised in Ireland and Scandinavia. **Flowers** July–September. **In gardens** Water up to 60cm deep, in full sun. Wildlife pools and fringes of larger lakes. **Image size** Half life-size.

When is a water lily not a water lily? When it is the fringed water lily and belongs to the bogbean family. This plant has water-lily-like mid-green leaves, occasionally blotched with purple and purple underneath. The sulphur-yellow flowers are funnel-shaped, with five fringed petals and are held above the surface on robust stems. Only one 3–4cm flower opens fully at a time. Once pollinated, its stem curves, pulling the flower underwater; the egg-shaped fruit rises to the surface only when the seeds are ripe and ready for dispersal.

Ragged robin [1]

Lychnis flos-cuculi (Caryophyllaceae)

Perennial. Height Up to 75cm. **Habitat** Damp meadows, wet woods and marshes. **Distribution** All northern Europe. **Flowers** May–June. **In gardens** Moist soil or shallow water in sun or shade. Bog gardens and pool margins. **Image size** Two-thirds life-size.

Ragged robin is hard to mistake, its flowers looking as if the petals have been shredded to thin segments, on slender stems with narrow opposite leaves. Country girls used to pick several unopened flowers, give each the name of a local youth, and keep them under her apron; the first to open would tell her who was destined to be her husband. Other country names include bachelor's buttons, referring to cloth buttons, and thunder-flower, because storms would follow if the plant was picked. *Flos-cuculi* means 'cuckoo-flower', a name used for plants that flower when cuckoos arrive in spring.

Red campion [2]

Silene dioica (Caryophyllaceae)

Perennial. Height Up to 1m. **Habitat** Woodland, hedges and waste ground; sometimes more open habitats on mountains. **Distribution** All northern Europe; introduced in Iceland. **Flowers** May–July. **In gardens** Any well-drained soil in sun or partial shade. Rock and woodland gardens. **Image size** Two-thirds life-size.

The red counterpart of white campion (p.40), this attractive hairy plant brightens up hedgerows and woodland edges. It can be identified by its clump-forming habit, with oblong to oval leaves – the lower stalked, the upper clasping the stem. The flowers have deeply notched petals, which are red or pink. They are either male or female, so two plants are needed to produce seed. Red and white campion are very promiscuous where they grow together. They cross freely, and the hybrid, *S.* x *hampeana*, is fertile, unlike most. This may hybridise with pure red campion, resulting in a swarm of hybrids with flowers in all shades of pink.

Thrift, sea pink [3]

Armeria maritima (Plumbaginaceae)

Perennial. Height Up to 30cm. **Habitat** Grassland, salt marshes, beaches, rocks and cliffs around coast; sometimes inland on mountains. **Distribution** Throughout northern Europe. **Flowers** April–July. **In gardens** Any well-drained soil in sun. Rock gardens and seaside plantings. **Image size** Two-thirds life-size.

Thrift used to appear on Britain's pre-decimalisation 12-sided threepenny piece – aptly, for this was a coin much saved in piggy banks. In fact its common name refers to its ability to thrive in situations where other plants will not. It has long roots to seek out water in dry conditions and can tolerate high concentrations of salt in the soil. It is attractive, with rosettes of grey-green linear leaves forming dense cushions, and dense pink (or sometimes white or almost crimson) fragrant flower heads, 15–25mm across, held high on hairy, leafless stems.

Marsh valerian [4]

Valeriana dioica (Valerianaceae)

Creeping perennial. Height Up to 40cm. **Habitat** Marshes, fens, bogs and wet meadows. **Distribution** All northern Europe except Ireland, Faeroe Is, Iceland and northern Scandinavia. **Flowers** May–June. **Image size** Two-thirds life-size.

A short, mostly hairless plant, marsh valerian spreads by means of stolons (creeping stems). Its basal leaves are long-stalked and entire; the stem leaves are pinnately lobed, with the individual leaflets toothed. The small five-petalled pale pink flowers are borne in dense terminal heads, with the male and female flowers on different plants. The males are bigger and lighter pink than the females, giving a slightly more open look to the male heads. The wind disperses the seeds on a feathery parachute of hairs. Its roots are said to smell unpleasantly, but not so badly as those of common valerian (p.159).

Butterbur [5]

Petasites hybridus (Asteraceae or Compositae)

Perennial. Height Up to 1.5m. **Habitat** Damp grassland, roadsides, and stream- and river-banks. **Distribution** Northern Europe as far north as Britain, Holland and Germany. Naturalised farther north. **Flowers** March–May. **Image size** Half life-size.

Butterbur is a plant of two halves. First the pinkish flowering spikes, with leaf-like scales and a dense mass of flower heads at the top, grow to a height of 30cm. Each head is made up either of female disc florets, which form parachute seeds, or of shorter-stalked male florets. Male and female plants are not always found together except in certain areas. Then come the leaves, which can be enormous – up to 1m across – and look rather out of place in normal surroundings. The leaves were once used to wrap butter, and the powdered root to remove skin blemishes.

Winter heliotrope, fragrant butterbur

[6] *Petasites fragrans* (Asteraceae or Compositae)

Evergreen perennial. Height Up to 40cm. **Habitat** Waste ground, waysides and stream-banks. **Distribution** Introduced. Naturalised in Britain, Ireland, Belgium, Holland, France and Denmark. **Flowers** January–March. **In gardens** Moist or wet soil in shade. Wild gardens and pondsides. **Image size** Half life-size.

An early-flowering escapee from gardens, this plant has large leaves up to 30cm across, which last through the winter. The flowering stems have up to 10 strongly scented flower heads, each consisting of lilac disc florets. The females have florets with short rays, but only male plants are found in Britain and probably also nearby parts of mainland Europe. It is an invasive plant that has succeeded not only in escaping from cultivation, but in spreading far and wide – probably by the regeneration of small pieces of rhizome that have broken off.

Purple saxifrage [1]

Saxifraga oppositifolia (Saxifragaceae)

Prostrate perennial. Height Up to 5cm. **Habitat** Damp rocky habitats and screes in mountains, often on limestone. **Distribution** Most of northern Europe. **Flowers** March–May. **In gardens** Well-drained neutral to alkaline soil in full sun. Rock gardens and alpine troughs. **Image size** Half life-size.

The very low-growing purple saxifrage forms glorious pale pink to purple blankets of colour wherever it hugs rocky ledges or screes and hangs in swathes off vertical cliffs; once seen, it is never forgotten. The voluptuous, almost stalkless, cup-shaped 1–2cm flowers have five petals and bluish anthers; they are held barely above the mats of tiny oval, greyish-green leaves. Long ago appreciated for its beauty, purple saxifrage was plucked from its mountainous home and transported to London's Covent Garden flower market; it remains a popular rock-garden plant.

Pink purslane [2]

Claytonia sibirica (syn. *Montia sibirica*; Portulacaceae)

Annual or short-lived perennial. Height Up to 40cm. **Habitat** Damp woods and stream-banks. **Distribution** Introduction, scattered throughout north-western Europe. **Flowers** April–July. **Image size** Half life-size.

A plant of western North America and north-east Asia, pink purslane has been grown in gardens since the late 18th century. It escaped almost 200 years ago, and has spread rapidly, colonising damp woodland and often forming impenetrable cover that chokes native vegetation. It does particularly well in areas, such as parts of the western and south-western British Isles, where conditions are very like the Pacific north-west of America. A hairless plant, it has basal leaves that are long-stalked, oval and pointed. The stem leaves are stalkless and grow in pairs. The pink flowers have five notched petals and two sepals; they grow in loose clusters.

Bearberry [3]

Arctostaphylos uva-ursi (Ericaceae)

Evergreen sub-shrub. Height Stems up to 1.5m long. **Habitat** Heaths, moorland and open woods on acid soil. **Distribution** All northern Europe except Faeroe Is. **Flowers** May–July. **In gardens** Any well-drained acid soil in full sun or partial shade. Ground cover and rock gardens. **Image size** Life-size.

A low evergreen shrub, bearberry has trailing stems that root at the nodes to forms a dense mat. Like many upland shrubs, it has leathery leaves covered with a thick shiny cuticle that prevents water loss in high winds. They are oval, widest at the outer end, dark green on top, and paler underneath. The 5–6mm drooping bell- or pitcher-shaped flowers are greenish-white with a pink tinge; they are formed of five sepals fused together, and grow from the leaf axils at the stem tips. Red berries, 6–8mm across, follow. These are tart but edible, and seem to be as effective as cranberries in treating urinary infections. But beware – the rest of the plant is poisonous.

Field madder [4]

Sherardia arvensis (Rubiaceae)

Annual. Height Up to 40cm. **Habitat** Cultivated and waste ground. **Distribution** All northern Europe except far north. **Flowers** May–October. **In gardens** Occasional lawn weed. **Image size** Life-size.

A sprawling hairless plant, field madder has whorls of narrowly elliptical or oval leaves. These are liberally endowed with small spines, which allow the plant to haul itself up through long grass. The tiny flowers are purplish-pink and four-lobed, with a long funnel-shaped tube formed by the fused petals. They are borne in dense clusters of up to 10 at the end of the stems. This is another plant that has undergone a decline in parts of its range, where grassland has been 'improved', leading to a healthy production of grass and a lack of what used to be common weeds.

Shining crane's-bill [5]

Geranium lucidum (Geraniaceae)

Annual. Height Up to 40cm. **Habitat** Shaded rocky areas, hedge-banks and walls. **Distribution** Most of northern Europe except Iceland and Faeroe Is. **Flowers** May–August. **In gardens** Liable to become a weed. **Image size** Life-size.

Although an elegant plant, shining crane's-bill seems to be bent on world domination. The glossy – 'shining' – green leaves, cut into five lobes and growing on stems that are easily broken, look the model of a well-behaved plant. Even the 15mm pink flowers, with un-notched petals and bristle-tipped sepals, look quite innocent. However, the beak-shaped seedpods' explosive mechanism for hurling the seeds far and wide is astonishing. The real shock comes the next year, when a single plant has managed to produce so many offspring. It is now a weed, appearing everywhere. In dry sunny weather, the plant turns red, to protect its green chlorophyll from the sun.

Bilberry, whortleberry, whinberry [6]

Vaccinium myrtillus (Ericaceae)

Dwarf shrub. Height Up to 60cm. **Habitat** Acid moorland, heaths and open woodland on poor soils. **Distribution** All northern Europe. **Flowers** July–September. **In gardens** Moist, acid, peaty or sandy soil in sun. Peat beds and tubs. **Image size** Half life-size.

Bilberry's blue-black berries – also called hurts or whorts – are smaller and more richly flavoured than commercial blueberries. They are rich in vitamins C and D, and are delicious eaten straight from the bush or made into crumbles, pies or jams. The Scots used them to produce a purple dye. It is a deciduous creeping shrub, with many upright branches, that forms dense mats. Its conspicuously veined oval, toothed leaves are bright green. The pink or greenish flowers droop singly or in pairs from the leaf axils; they are bell-shaped and 4–6mm long.

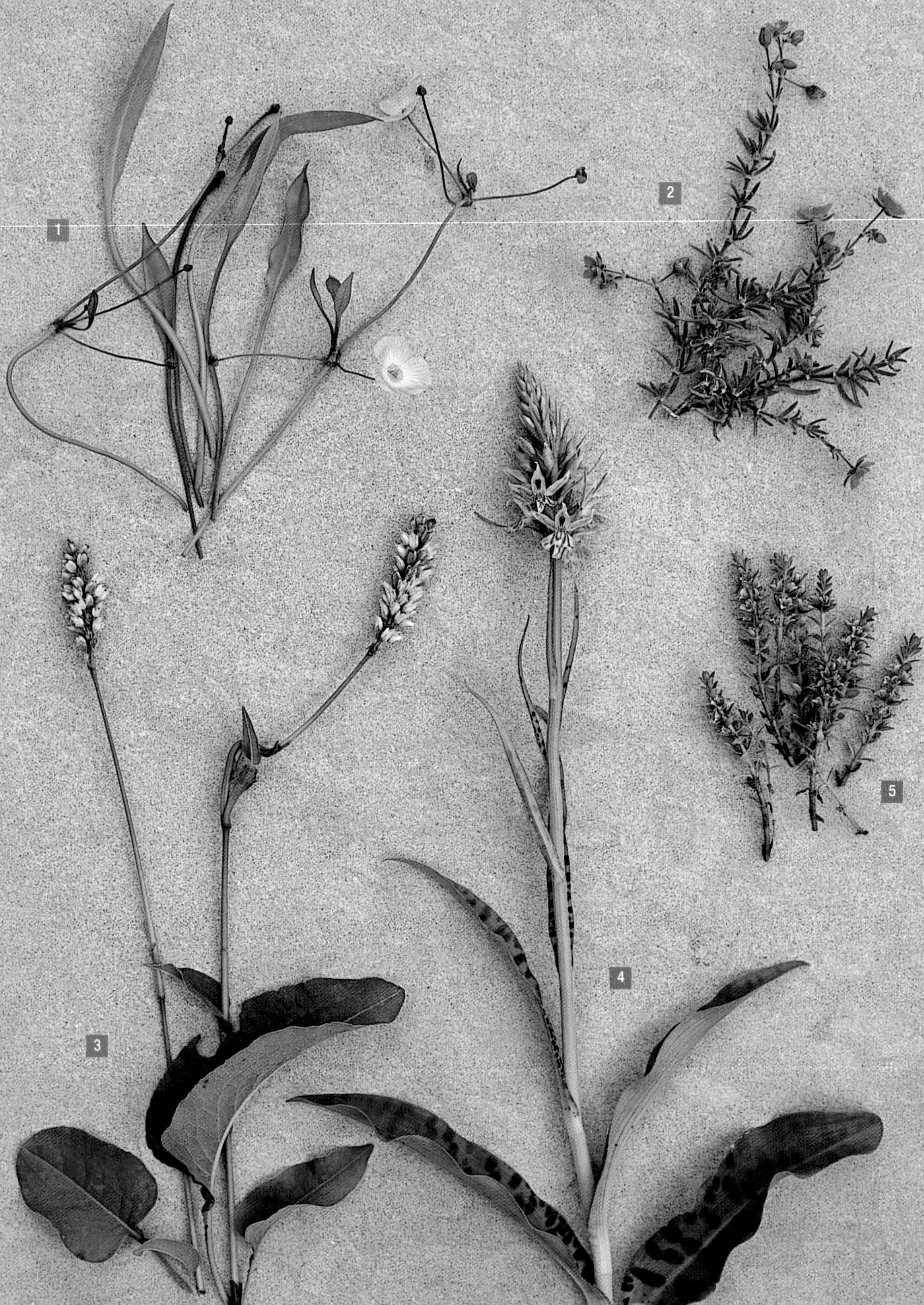
1
2
3
4
5

Lesser water plantain [1]
Baldellia ranunculoides (Alismataceae)

Semi-aquatic perennial. Height Up to 20cm.
Habitat Ponds, ditches and streamsides. **Distribution**
Northern Europe as far north as southern Norway.
Flowers May–August. **In gardens** Wet soil or shallow
water. Pool margins. **Image size** Life-size.

Not related to true plantains, this plant has
a rosette of narrowly lanceolate, plantain-
like leaves. The 10–15mm flowers grow in
whorls, with three green sepals and three
pale mauve-pink to whitish petals, each
with a splash of yellow at the base. Lesser
water plantain is sensitive to changes in the
water. Run-off leaching nitrate fertiliser
from local fields causes excessive growth
that chokes ponds, while drainage of
wetlands can destroy its habitat.

Rock sea spurrey [2]
Spergularia rupicola (Caryophyllaceae)

Perennial. Height Up to 20cm. **Habitat** Coastal cliffs,
rocks and walls. **Distribution** Britain, France and Ireland.
Flowers June–September. **Image size** Life-size.

The woody stock of rock sea spurry produces
a number of purple stems covered with
sticky hairs. Several fleshy leaves – parallel-
sided and flattened, with a horny tip –
grow at each node. The flowers are borne in
a loose inflorescence of up to 20 individuals.
Each is 8–10mm across and deep pink in
colour. Once the flower is fertilised, the
stalk bends back, gradually straightening as
the seeds ripen. Rock sea spurrey is one of a
number of plants that live happily where
others are scorched or washed off.

Bistort, snakeweed [3]
Persicaria bistorta (syn. *Polygonum bistorta*;
Polygonaceae)

Perennial. Height Up to 80cm. **Habitat** Damp grassy
places and wasteland. **Distribution** All northern Europe
except Finland, Iceland, Faeroe Is. **Flowers** May–August.
In gardens Moist, well-drained soil in sun or partial shade.
Mixed borders and beside ponds. **Image size** Life-size.

The species name *bistorta* is Latin for
'twice-twisted'. It refers to the plant's
contorted rootstock, which enables it to
spread to form substantial clumps and is
also the origin of the country name
snakeweed. Thin, unbranched stems grow
from it, with oval alternate leaves up to
20cm long. Where the basal leaves meet the
stem is a papery sheath; the upper leaves
have no stalk and clasp the stem. At the tip
of each stem grows a crowded flower spike,
up to 9cm long, consisting of numerous
tiny pink paired male and female flowers.
In northern England the leaves are still
made into a Lenten pudding, Easter ledge.

Common spotted orchid [4]
Dactylorhiza fuchsii (Orchidaceae)

Tuberous perennial. Height Up to 60cm. **Habitat** Open
woods, fens, grassland and other open areas on chalk or
limestone. **Distribution** All northern Europe except
Iceland and Faeroe Is. **Flowers** May–August. **In gardens**
Moist, well-drained humus-rich soil in partial shade.
Woodland gardens. **Image size** Life-size.

This is likely to be the most common orchid
in the British Isles and many other parts of
northern Europe. It is easily recognised by its
blunt basal leaves, which have more
elongated spots than other species. The
flowers – which grow in dense spikes – have
a very well defined lip, which is three-lobed,
with the smallest but longest central lobe a
narrow triangle. Common spotted orchid is
very variable, with flowers ranging from
reddish-purple through pink to white. It
grows to less than 30cm on chalk grassland,
but may be up to twice as high in damp
meadows. It is also promiscuous, hybridising
with at least six other species of *Dactylorhiza*.

Sea milkwort [5]
Glaux maritima (Primulaceae)

Creeping perennial. Height Stems up to 30cm long.
Habitat Coastal rocks, sand, mud and salt marshes.
Distribution All northern Europe except Faeroe Is.
Flowers June–August. **Image size** Life-size.

This is another species that has adapted to
very salty conditions. Water uptake is very
difficult, so sea milkwort uses its fleshy
leaves as water-storage organs, with a thick,
grey-green cuticle to prevent water loss.
The plant also roots at the nodes, forming a
dense mat that helps to protect it against
drying by the wind. The small, stalkless,
pale pink to purple (or sometimes white)
bell-shaped flowers are solitary and grow in
the junction of leaf and stem. They have no
petals, but the five sepals are coloured.

Cuckoo flower, lady's smock [6]
Cardamine pratensis (Brassicaceae or Cruciferae)

Perennial. Height Up to 60cm. **Habitat** Damp grassland
and waterside habitats. **Distribution** Throughout
Northern Europe except Iceland. **Flowers** April–June.
In gardens Moist, humus-rich soil in a sunny position.
Mixed borders. **Image size** Life-size.

This plant has pinnate leaves consisting of
many leaflets, the lower ones in a rosette
and rounder than the narrow stem leaflets.
The flowers – that come with the cuckoo –
range from lilac-pink to white and are borne
in a loose inflorescence. The fruits disperse
the seeds explosively. Cuckoo flower is very
variable and its variable chromosome count
indicates that it is a species in transition.

Bee orchid [1]

Ophrys apifera (Orchidaceae)

Perennial. Height Up to 50cm. **Habitat** Grassland, field edges, banks and dunes, especially on chalk or limestone. **Distribution** Britain, Ireland, Belgium, France, Germany. **Flowers** June–July. **Image size** Three times life-size.

This orchid is aptly named. At first sight it seems to have a bumblebee sitting on each flower. Closer inspection reveals the 'bee' to be the flower's lower lip. The plant has set a sexual trap: the lip is an exact replica of the female of a European bumblebee. The flower emits the scent of female bee pheromones, which attract males. A male bee will try to mate with this 'female' and, while it is clasping the lip, pollen masses attach to its head. Eventually, the frustrated bee detaches itself, and moves on to be duped again by another flower, carrying the pollen to the stigma as it goes. Despite this strategem, most British bee orchids are self-pollinated and their numbers fluctuate year on year.

Fragrant orchid [2]

Gymnadenia conopsea (Orchidaceae)

Perennial. Height Up to 45cm. **Habitat** Bogs, fens, scrub and chalk or limestone grassland. **Distribution** All northern Europe except Iceland and Faeroe Is. **Flowers** June–August. **Image size** One-third life-size.

Living up to their name, a group of fragrant orchid plants together will, on a still day, fill an area with a perfume comparable to the clove scent of carnations. The flowers grow in a dense cylindrical spike, and may be various shades between red-purple and pale lilac-pink or occasionally white. The fragrant orchid has spreading outer petals, and a lip with a long tapering spur. They are pollinated by long-tongued moths and butterflies. The leaves are long, narrow and keeled; they are hooded at the tip. Those near the top of the stem are much smaller and bract-like. Fragrant orchid numbers vary enormously from year to year.

Early marsh orchid [3]

Dactylorhiza incarnata (Orchidaceae)

Tuberous perennial. Height Up to 40cm. **Habitat** Marshes and damp grassy habitats. **Distribution** Most of northern Europe except Iceland and far north. **Flowers** May–July. **Image size** Two-thirds life-size.

Land drainage is causing a slow decline of the early marsh orchid, a pretty plant of damp habitats and one of 25 marsh orchid species. The upright flower spikes bear pinkish or lilac (or sometimes even whitish, yellow or red) flowers whose upper sepals and petals are fused together to form a hood between two lateral sepals. The lower lip is delicately spotted and the sides are folded back to give it a narrow appearance. The leaves are yellowish-green, lanceolate, keeled and usually – unlike those of many orchids – without spots. There are, however, at least five subspecies and one, *D. incarnata* subsp. *cruenta* (*D. cruenta*), does have spotted leaves.

Pyramidal orchid [4]

Anacamptis pyramidalis (Orchidaceae)

Perennial. Height Up to 45cm. **Habitat** Grassland and woodland clearings on chalk and limestone, and dunes. **Distribution** All northern Europe except Iceland, Faeroe Is, Norway, northern Sweden and Finland. **Flowers** June–August. **Image size** One-and-a-half life-size.

Pyramidal orchid is relatively common, and is one of the last summer grassland orchids to flower. It may grow in large numbers, the bright magenta to pale pink pyramid-shaped flower spikes enlivening midsummer meadows. Each flower has six petals and sepals, all the same colour. The inner bottom segment forms a three-lobed lip and has a long backward-projecting spur, which stores nectar. The plant's fragrance is best described as fox-like, but pollinating insects seem to love it. As the flower head ages, it lengthens, losing the pyramidal shape and becoming conical. The colour also fades slightly.

Sea bindweed [5]

Calystegia soldanella (Convolvulaceae)

Creeping and trailing perennial. Height Stems up to 60cm long. **Habitat** Coastal sand, shingle and dunes. **Distribution** Coasts of northern Europe except Iceland, Faeroe Is and Scandinavia. **Flowers** June–August. **Image size** Life-size.

At first sight, this is field bindweed (see below) by the sea. The funnel-shaped flowers have the same pink and white colour scheme, but closer inspection reveals that those of sea bindweed are at least twice as big. The stems twine much more weakly, and a quick look at the leaves is conclusive. Sea bindweed has kidney-shaped, slightly fleshy leaves, with the leaf-stalk longer than the leaf blade. It grows from a rhizome which creeps through sand or shingle, providing both anchorage and a means of gathering as much fresh water as possible.

Field or common bindweed [6]

Convolvulus arvensis (Convolvulaceae)

Twining perennial. Height Stems up to 2m long. **Habitat** Wasteland, hedgerows, arable land and gardens. **Distribution** All northern Europe except Iceland and Faeroe Is. **Flowers** June–September. **In gardens** An almost universal weed. **Image size** Two-thirds life-size.

With delicate pink funnel-shaped flowers up to 25mm across, whose five narrow white stripes are set off by five broad mauve stripes on the underside, bindweed looks an elegant plant. The stems, however, reveal its true nature. They trail on the ground at first but, once on to anything that points upwards, they are away. They coil anti-clockwise around the support and strangle any relatively soft-stemmed plant. Growth is so fast that a stem can complete one coil in less than two hours. The triangular or arrow-shaped leaves are borne alternately up the stem, with flowers in the upper axils. Bindweed has deep roots, which are quite easy to break, so any attempt to remove it always leaves a piece behind.

Martagon or Turk's-cap lily [1]

Lilium martagon (Liliaceae)

Bulbous perennial. Height Up to 1.5m. **Habitat** Woods, scrubland and mountain grassland. **Distribution** All except far northern Europe but native only in the south. **Flowers** June–August. **In gardens** Deep, humus-rich soil in dappled shade; plant bulb deeply. Woodland gardens, mixed borders or naturalise in grass. **Image size** Two-thirds life-size.

This native of the mountains of Europe has been grown in gardens since the 1590s, so it is no surprise that it has escaped from cultivation. It seems to be the very model of a native plant and there are certainly many uglier garden escapees. Martagon lily has a graceful red-spotted stem with whorls of shiny dark green leaves. It is topped by a cluster of nodding flowers, each with six pink to purple tepals, which are curved back to reveal the pinkish stamens and style. The name comes from the French *martagan*, the name for a type of turban, from the flower shape – hence the name Turk's-cap lily.

Sainfoin [2]

Onobrychis viciifolia (Fabaceae or Leguminosae)

Perennial. Height Up to 80cm. **Habitat** Grassy, cultivated and waste ground, especially on chalky soil. **Distribution** Throughout most of northern Europe, but often naturalised. **Flowers** June–August. **Image size** Two-thirds life-size.

Sainfoin was once thought to be named after a St Foyne – who was found to be apocryphal. More prosaically, the name is from French words meaning 'wholesome hay', indicating its value as a fodder crop. It is an upright, often branched plant, with leaves consisting of up to 14 pairs of narrow leaflets, plus a single leaflet at the end. It has showy pink or red flowering spikes, and can look very much part of the landscape. In such places, especially if on chalk, it may be a native or at least a very long-standing introduction. However, it is usually a fodder plant that has escaped to a life of freedom.

Grass vetchling [3]

Lathyrus nissolia (Fabaceae)

Annual. Height Up to 90cm. **Habitat** Grassy areas on chalk or alkaline clay. **Distribution** Most of northern Europe except Ireland, Iceland and Scandinavia. **Flowers** May–July. **Image size** Two-thirds life-size.

Grass vetchling usually manifests itself as a magenta pea flower apparently hanging in mid-air. Closer scrutiny reveals that the flower has a long thin stalk, anchoring it to what looks uncannily like a grass plant. This is actually a modified leaf and leaf-stalk, reduced to what resembles a blade of grass. Once the long 3–6cm seedpods form, the sight of one dangling on a grass-like stem is even more perplexing. Grass vetchling, like many annuals, has seeds that can lay dormant for a considerable period of time. Disturbing the soil may often bring on an unexpected crop of this intriguing plant.

Maiden pink [4]

Dianthus deltoides (Caryophyllaceae)

Perennial. Height Up to 45cm. **Habitat** Dry, usually sandy or chalky grassland. **Distribution** Throughout northern Europe, except Ireland, Faeroe Is and Iceland. **Flowers** June–September. **In gardens** Well-drained soil in full sun. Rock gardens, raised beds and front of borders. **Image size** Two-thirds life-size.

A loosely tufted plant, maiden pink has short creeping sterile shoots and flowering stems that form open clusters. The flowers are up to 20mm across, white to cerise-pink and usually spotted, and lack any scent. The plant's common name has been said lyrically to come from the pink flush on a maiden's cheek, but there is a more down-to-earth version that says it is a corruption of 'mead (or meadow) pink'. Maiden pink depends almost entirely on its habitat for survival, which has been on the decline until recently. The realisation that once a habitat has gone the species has gone, combined with some timely EU directives, may allow this plant to survive for longer.

Bird's-eye primrose [5]

Primula farinosa (Primulaceae)

Perennial. Height Up to 15cm. **Habitat** Damp grassland, rocks and peaty areas, on limestone. **Distribution** Northern England, Scotland and continental Europe as far north as Denmark and southern Sweden. **Flowers** May–July. **In gardens** Very well-drained humus-rich soil in sun. Raised beds and pots. **Image size** Two-thirds life-size.

The species name *farinosa* means 'floury', and the underside of the leaves and the young stems look as though they have been covered in fine flour. The leaves grow in a rosette and are broader above the middle, gradually tapering into the base. The flower spike ends in a cluster of two to five lilac to rose-pink flowers, each of which has a yellow central eye. Bird's-eye primrose is very much a plant of wild open spaces. It may be abundant where it grows, but its habitats are fragile, and need special care. If there are a few unfortunate seasons or surrounding areas are treated inappropriately, the plant could face a bleak future.

Sweet briar [6]

Rosa rubiginosa (syn. *R. eglanteria*; Rosaceae)

Shrub. Height Up to 2.5m. **Habitat** Grassy and coastal habitats, open copses, scrub and banks. **Distribution** All northern Europe except Faeroe Is and Iceland. **Flowers** June–July. **In gardens** Moist, well-drained soil, full sun. Hedges, shrub borders. **Image size** Two-thirds life-size.

The strong apple scent that exudes from the foliage of the sweet briar, when crushed or after a heavy rain shower, is a unique characteristic of this wild rose, and has endeared it to gardeners for centuries. On the other hand, the deep pink flowers surprisingly have very little scent. The stems carry both hooked thorns and bristles. The leaves consist of five to seven yellowish-green leaflets, tinged a rusty shade. Almost elliptical bright red hips appear in late summer and early autumn. Shakespeare and other poets immortalised the sweet briar as the 'eglantine rose'.

Mind-your-own-business [1]
Soleirolia soleirolii (syn. *Helxine soleirolii*; Urticaceae)

Mat-forming perennial. Height Up to 10cm. **Habitat** Damp shady areas, and on walls. **Distribution** Introduced. Britain, Ireland, Belgium, France, Holland. **Flowers** May–October. **In gardens** Any damp soil in a shady position. Ground cover. **Image size** One-and-a-half life-size.

A western Mediterranean native, this small mat-forming plant was probably a weed of glasshouses before it crept outside. It has proved to be remarkably resilient and hardy in any suitable habitat. Very different from the stinging nettle (p.218) of the same family, it has tiny (5mm) round leaves, and is so called because a probing finger leaves a permanent telltale depression in the mound of foliage. Its other common names include baby's tears and mother-of-thousands. The leaves arise from very slender stems, which root at the nodes. The inconspicuous pink flowers grow singly in the leaf axils. Each flower is either male or female, but both are found on the same plant.

Dog rose [2]
Rosa canina (Rosaceae)

Scrambling shrub. Height Up to 3m or more. **Habitat** Scrub, woods and hedge-banks. **Distribution** All northern Europe except Iceland and Faeroe Is. **Flowers** June–August. **In gardens** Any fertile, well-drained soil in a sunny position. Hedges and scrambling into trees in wild gardens. **Image size** Life-size.

It's not clear why such a handsome rose should attract the unflattering epithet 'dog'. It has been said that an extract of its roots was used by the Romans to treat rabies, or that its thorns look like canine teeth; it is not – unlike other 'dog' plants – unscented. The pink or white flowers are up to 5cm across. The leaves have up to three pairs of leaflets, plus one at the tip. The fruits – hips or heps – eaten by animals and birds, are used to make wine and liqueur and are the main ingredient of rose-hip syrup, a rich source of vitamin C. Dog rose is very variable.

Lousewort [3]
Pedicularis sylvatica (Scrophulariaceae)

Perennial. Height Up to 25cm. **Habitat** Bogs, marshes, and damp heathland and open woods. **Distribution** All northern Europe except Faeroe Is, Iceland and the far north. **Flowers** April–July. **Image size** Two-thirds life-size.

The name comes from the plant's reputation, in both Britain and Germany, of giving lice to livestock. This probably arose because lousewort flowers when louse infestation is increasing. However, the plant may well carry the snails that harbour liver-fluke larvae, which sheep take in as they graze. Liver flukes can be fatal and any sheep with an infestation is also likely to carry lice. Lousewort is a hemi- (or semi-) parasite, taking water and mineral salts from plant roots. It is hairless, and has more or less upright stems growing from the base. The fern-like leaves are oblong and deeply divided into leaflet-like lobes, with toothed divisions. The pink or white flowers are borne in spikes at the end of the branches. Each flower is up to 25mm long and two-lipped. The fused sepals form an angled tube when in flower, with four small toothed lobes.

Common vetch [4]
Vicia sativa (Fabaceae or Leguminosae)

Trailing or climbing annual. Height Up to 1.5m. **Habitat** Grassland, hedges, roadsides and cultivated ground. **Distribution** All northern Europe, but introduced in Iceland and Faeroe Is. **Flowers** May–September. **Image size** Life-size.

Common vetch either sprawls or climbs, depending on how strong the surrounding plants are. The leaves have four to eight pairs of leaflets, which end in a branched tendril. The 18–30mm flowers are borne – singly or in pairs – on long stalks. They are usually pink or pink-purple, but some plants may have white or bicoloured flowers. The usually hairy seedpods contain up to 12 seeds. Common vetch is a variable plant. The British native form has smaller flowers

and narrower leaflets than those elsewhere. The cultivated form has larger flowers and is more robust.

Bird's-foot [5]
Ornithopus perpusillus (Fabaceae or Leguminosae)

Annual. Height Up to 30cm. **Habitat** Dry, sandy or gravelly areas – bare or grassy. **Distribution** Northern Europe to S. Sweden except Norway. **Flowers** May–August. **Image size** One-and-a-third life-size.

Although it is related to the bird's-foot trefoils (pp.91 & 100) and has similar curved and pointed seedpods that resemble birds' claws, bird's-foot has quite different leaves. They consist of up to 12 pairs of elliptical leaflets, with a single leaflet at the tip. These also distinguish it from the pink-flowered (though taller-growing) common vetch (see left), which has a tendril rather than a terminal leaflet. Bird's-foot is a downy, often sprawling plant. The flower heads have long stalks, arising from a leaf axil, and have a leaf-like bract underneath. Each has up to six small whitish or pink flowers with red veins. These normally self-pollinate, producing a pod that is divided into segments – one seed to each. The pod breaks up to release the seeds.

Cut-leaved deadnettle [6]
Lamium hybridum (Lamiaceae or Labiatae)

Annual. Height Up to 45cm. **Habitat** A weed of waste, disturbed and cultivated ground. **Distribution** All northern Europe. **Flowers** March–October. **Image size** One-and-a-half life-size.

Very similar to red deadnettle (*Lamium purpureum*; p.144) and often growing with it, cut-leaved deadnettle is usually less downy-hairy and generally more slender. The leaves are more deeply cut, with longer teeth. The flowers are pink-purple, often with a shorter corolla tube. It is probably a hybrid between red deadnettle and an Aegean species, *L. moschatum*. Unlike most hybrids, it is fertile.

Common centaury [1]

Centaurium erythraea (Gentianaceae)

Biennial. Height Up to 50cm. **Habitat** Dry grassland, scrub and dunes. **Distribution** All northern Europe except Faeroe Is, Iceland, Norway and Finland. **Flowers** June–September. **Image size** Life-size.

Common centaury is hairless, with a rosette of elliptical leaves and shorter stem leaves, topped by clusters of pink or purplish five-petalled 10–15mm flowers. However it is a very variable plant. In good conditions it is much branched, with many flower heads but, exposed to wind and rain, it can form a mound only a few centimetres high. As a result, some very tiny plants have been put into different species. Centuary has been used as a herbal cure-all for everything from wounds inflicted by the many-headed Hydra of Greek mythology (Chiron the centaur used it in this way) to diseases and even freckles.

Rosebay willowherb [2]

Chamerion angustifolium (syn. *Epilobium angustifolium*; Onagraceae)

Spreading perennial. Height Up to 1.5m. **Habitat** Waste ground, open woods and mountain screes. **Distribution** All northern Europe. **Flowers** June–September. **In gardens** Any soil in sun. Wild gardens, but may become a weed. **Image size** Life-size.

The Victorians regarded rosebay willowherb as a fine plant but somewhat difficult to cultivate. It has clearly picked up a great deal of fighting spirit since then. It became known as 'fireweed' after it colonised bomb sites during World War II, and it has not looked back since – often to the detriment of gardens. (It seeds itself freely and also spreads by means of rhizomes.) Its long spikes of pink, four-petalled 20–30mm flowers – with the lower pair of petals larger than the upper – and its alternate leaves shaped like willow leaves, make it easily recognisable. The long, four-angled seedpods release seeds that have an attached plume, like those of willow seeds.

Great willowherb [3]

Epilobium hirsutum (Onagraceae)

Spreading perennial. Height Up to 1.6m. **Habitat** Fens, marshes, streamsides and other damp places. **Distribution** All northern Europe except Faeroe Is and Iceland. **Flowers** July–September. **In gardens** Any damp soil in sun. Wild gardens, but may spread vigorously. **Image size** Life-size.

A great show-off of a plant, with hairy stems and opposite stalkless leaves, great willowherb has bright purple-pink 15–25mm flowers with four notched petals all the same size. They are followed by long narrow fruits full of plumed seeds. Like rosebay willowherb (see left), it spreads by both seeds and creeping rhizomes. Great willowherb has many local names, such as apple pie and codlings and cream. This last is probably an old reference to the 'codde', or young capsule beneath the flower, than to apples.

Deptford pink [4]

Dianthus armeria (Caryophyllaceae)

Annual or biennial. Height Up to 50cm. **Habitat** Dry grassy places on sandy or chalky soil. **Distribution** All northern Europe except Iceland, Faeroe Is, Norway, northern Sweden and Finland. **Flowers** June–August. **In gardens** Well-drained soil in sun. Rock gardens. **Image size** Life-size.

Deptford pink is misnamed. It was said by 16th-century herbalist John Gerard to be from 'the great field next to Detford', but he then gaven an accurate description of maiden pink (p.135). Unfortunately the mistake was perpetuated, and *Dianthus armeria* and Deptford – now part of south-east London – are linked forever. The plant is dark green, with a rosette of long thin basal leaves and paired stem leaves. The unscented starry pink flowers, 10–15mm across, are held in dense clusters surrounded by linear bracts. It is a plant that, thanks to loss of its habitat, is losing the race for survival in the wild. Seedlings and seed are available from specialist nurseries.

Bramble, blackberry [5]

Rubus fruticosus agg. (Rosaceae)

Scrambling, spreading sub-shrub. Height Up to 3m. **Habitat** Woods, scrub, hedges, roadsides and banks. **Distribution** All northern Europe. **Flowers** May–September. **In gardens** Any soil in sun or part-shade. Fruit and wild gardens. **Image size** Half life-size.

The blackberry is the best example of a fine fruit well protected; few will extract the prize without the thorns exacting revenge. *Rubus fruticosus* is, in fact, an aggregate of something like 2,000 micro-species, which each keep their individual characteristics by reproducing without the need for sex, in a process called apomixis. All share basic characters, with leaves usually three- to five-lobed, vicious thorns, and the ability to cover huge areas of ground by growing an upright stem, which arches and roots where it touches the ground. Once each plant has done its best to take over the world, pink or white flowers appear, followed by the shiny sweet, black fruits. In any well-grown bramble patch, it is usually a race between humans, blackbirds, wood mice and others to get the fruit. Humans occasionally win.

Flowering rush [1]

Butomus umbellatus (Butomaceae)

Aquatic perennial. Height Up to 1.5m above water level. **Habitat** Margins of rivers, canals, lakes and ponds; also ditches. **Distribution** Many parts of northern Europe except Iceland and Faeroe Is. Often introduced. **Flowers** July–August. **In gardens** Fertile soil (best in a container) in water up to 25cm deep. Margins of any except the smallest pool. **Image size** Life-size.

Many plants growing in water have long, thin parallel-sided leaves; of these, some are greyish-green and some triangular in cross-section like the flowering rush. None, however, have this plant's elegant flower heads. These have flower stalks that grow from the same point on the stem, giving an umbrella-shaped flower head – an umbel. The flowers open in sequence, and are up to 25mm across, with three sepals and three pink petals. Although a native plant, flowering rush is also naturalised by well-meaning (but misguided) pond-owners releasing some of their excess into the wild.

Round-leaved crane's-bill [2]

Geranium rotundifolium (Geraniaceae)

Annual. Height Up to 40cm. **Habitat** Walls, hedge-banks, bare and grassy places. **Distribution** Northern Europe from Britain and Ireland, Belgium and Germany southwards. **Flowers** May–July. **Image size** Life-size.

A downy plant with occasional glandular hairs, round-leaved crane's-bill is usually branched from the base. The lower leaves are generally five-lobed, the upper smaller and often more deeply cut – less deeply than many other *Geranium* species, but hardly as round as the name suggests. The 12mm flowers grow on long stalks, with glandular hairs on the sepals; the pink petals are either rounded at the top or slightly indented. The beak-shaped fruits spread the seeds by an effective catapult action as they dry and split. Round-leaved geranium seems to be spreading in the wild, mainly on roadsides, railway ballast

and other easily invaded situations. Once in place, it seems to have staying power, for colonies tend to survive.

Musk mallow [3]

Malva moschata (Malvaceae)

Perennial. Height Up to 80cm. **Habitat** Hedge-banks, meadows, field margins, roadsides and other grassy places. **Distribution** Most of northern Europe except Iceland; introduced in Scandinavia. **Flowers** June–August. **In gardens** Any well-drained soil in full sun. Mixed and herbaceous borders, and wild and cottage gardens. **Image size** Two-thirds life-size.

Musk mallow has stiff, upright, hairy shoots. The basal leaves are long-stalked and kidney-shaped, but farther up the stem the stalks are shorter and the leaves are divided into narrow segments. The large, somewhat poppy-like, fragrant flowers are borne in the axils of the upper leaves; they are up to 6cm across, with the petals shallowly notched at the tips. They are usually bright to pale rose-pink but may be white. The fruits uncannily resemble a small cheese cut expertly into wedges, each wedge being an individual seed. These fruits have earned mallows the local name of cheese flower. Musk mallow has long been a popular cottage-garden plant.

Common mallow [4]

Malva sylvestris (Malvaceae)

Perennial. Height Up to 1m. **Habitat** Roadsides, rough ground, meadows and other grassy places. **Distribution** Northern Europe except Faroe Is, Iceland. **Flowers** June–September. **In gardens** Any well-drained soil in full sun. Mixed and herbaceous borders, and wild and cottage gardens. **Image size** Two-thirds life-size.

Common mallow is a robust, hairy plant with rounded lower leaves and stem-leaves that are distinctly five-lobed but less deeply cut than musk mallow's (see above). The flowers are crisper and up to 4cm across,

and are rose-pink to pinkish-purple, with dark veins. The narrow petals have a wide notch at the tip. The typical 'cheese' fruits have green-brown seeds. The young shoots were eaten as a vegetable in pre-Roman times and the whole plant was later used as an anti-aphrodisiac, to promote sobriety and calm. There are several garden cultivars in magenta, white and delicate blue.

Moss campion [5]

Silene acaulis (Caryophyllaceae)

Perennial. Height 3–10cm. **Habitat** Rocks, screes and short grass on mountains. **Distribution** All northern Europe except Belgium, Holland and Denmark. **Flowers** May–August. **In gardens** Well-drained fertile soil in sun; protect from winter wet. Sinks, rock gardens and alpine beds. **Image size** Half life-size.

Come across a patch of moss campion out of flowering time and it looks like a well-grown hummock of moss. It is ideally adapted to withstand the testing conditions on mountains. The bright green, small linear leaves form a perfect dome, deflecting the worst of the roughest winds. When the plant comes into flower, it reveals itself as a true gem, with cerise-pink flowers up to 1cm across on short stalks. Each flower has five notched petals. Moss campion is a true mountaineer, hardly ever being found below 600m and growing far up into the Arctic; on Alpine mountains, it has been found at heights of up to 3700m.

Common restharrow [1]

Ononis repens (Fabaceae or Leguminosae)

Perennial or sub-shrub. Height Up to 60cm. **Habitat**
Dry rough grassland and meadows on sandy soil,
especially by the sea. **Distribution** All northern Europe,
except Iceland, Faeroe Is and northern Scandinavia.
Introduced in Finland. **Flowers** June–September.
Image size Two-thirds life-size.

Restharrow is a mat-forming plant with
stems that are woody at the base and self-
rooting at the nodes. However, its singular
characteristic – the one that gave it its
common name – is its long, thick
underground rhizomes. In the days when
farm implements were horse-drawn, this
and the matted wiry stems would literally
'arrest the harrow', making it unpopular
with farmers. It was small compensation
that the roots were chewed by children as a
poor substitute for liquorice. It was also
unpopular for another reason: its foetid,
almost petroleum-like smell, which would
taint milk and butter if it was eaten by
cattle. Restharrow's leaves vary from a single
blade to three-lobed, and there may be the
occasional soft spine on the stem. The
pretty pink pea flowers, 15–20mm long,
grow in a loose inflorescence. They are
followed by short seed pods.

Hemp agrimony [2]

Eupatorium cannabinum (Asteraceae or Compositae)

Perennial. Height 1–1.5m. **Habitat** Streamsides and
marshy areas, and damp woods, wasteland and
grassland. **Distribution** All northern Europe except
Faeroe Is and Iceland. **Flowers** July–September.
In gardens Reasonably moist soil in sun. Informal
borders and wild gardens. **Image size** Two-thirds life-size.

A large, handsome plant with reddish hairy
stems, hemp agrimony has leaves that are
divided into five lobes, supposedly like a
hemp (*Cannabis*) leaf. Numerous dense,
broad flower heads grow at the ends of
branches, each consisting of many
individual flowers and circled by a ring of
purple-tinged leaf-like bracts. The flowers
have no ray florets but up to six whitish-
pink to reddish-mauve disc florets and
protruding stamens. The seeds have a
white parachute of hairs for wind dispersal.
The genus *Eupatorium* is named after
Eupater (King) Mithradates of Pontus
(d. 63BC), who used the plant to prepare
antidotes to poisons in an era when to be
royal was to be in danger and a wise ruler
took precautions.

Apple mint [3]

Mentha x *villosa* (Lamiaceae or Labiatae)

Perennial. Height Up to 1m. **Habitat** Rough ground and
wasteland. **Distribution** Probably throughout northern
Europe, but mainly as an introduction. **Flowers**
July–September. **In gardens** Any well-drained moist soil
in sun or partial shade. Best grown in a container.
Image size Two-thirds life-size.

This extremely variable hybrid depends
greatly on how much resemblance it
bears to its parents, round-leaved mint
(*Mentha suaveolens*; also sometimes known
as apple mint) and the pointed-leaved
spearmint (p.156). It may be hairy or
hairless, with more or less round to much
longer leaves – which are, as the name
suggests, apple-scented. The pink flower
heads are merely decorative – the flowers
are sterile. Any mint will try to take over a
garden, so plant it in a pot – free-standing
or sunk in the ground – and keep a close
eye on it.

Soapwort [4]

Saponaria officinalis (Caryophyllaceae)

Spreading perennial. Height 60–90cm. **Habitat** Grassy
places, woods, hedgerows, and waste and formerly
cultivated ground. **Distribution** Native or naturalised
throughout most of northern Europe except Iceland and
Faeroe Is. **Flowers** July–October. **In gardens** Good soil
in sun or dappled shade. Wild gardens and borders.
Image size Two-thirds life-size.

Soapwort makes a lather when boiled, and
was once used to wash the natural oils out
of wool. It was grown next to woollen mills
and made its way easily into the wild from
there, for it self-seeds freely. It has a
creeping stem, which produces hairless
upright flowering stems. The leaves are oval
or elliptical, with prominent veins. The pale
pink 25–35mm flowers grow in clusters;
they have a faint scent. Even in the wild, a
double-flowered form is quite common,
and probably gave the plant its other
common name, bouncing Bet. Several
garden cultivars are derived from it.
Soapwort also had many herbal uses,
despite containing toxic saponin.

Orpine [5]

Sedum telephium (Crassulaceae)

Perennial. Height Up to 60cm. **Habitat** Woods, hedges
and grass, usually on sandy soil. **Distribution** All
northern Europe except Iceland and Faeroe Is.
Flowers July–September. **In gardens** Well-drained soil.
Mixed borders. **Image size** Two-thirds life-size.

This exotic-looking succulent looks
completely out of place in woods or on
hedge-banks and, in many cases, it is
indeed a garden escapee. Yet it is a native
plant, and the grey-green fleshy leaves
reflect the fact that woodland can be a very
dry place to grow, thanks to the umbrella of
the tree canopy and competition from
roots. Big heads of rosy-red (or whitish,
greenish or yellowish) flowers come later in
the year, when rain is often not far off.
The strategy works – orpine is also called
'live long', and it does. In fact, orpine is a
misnomer; *or* is French for 'gold', and
applies to the related golden-yellow biting
stonecrop (p.96). Fine garden cultivars of
orpine include 'Atropurpureum',
flushedwith purple, and 'Variegatum', with
green, cream and pink foliage.

Bog pimpernel [1]

Anagallis tenella (Primulaceae)

Prostrate perennial. Height Stems up to 40cm long.
Habitat Bogs, marshes, peaty ground and wet grassland.
Distribution Britain, Ireland, Belgium, France, Holland
and Germany. **Flowers** June–August. **In gardens** Moist,
well-drained soil in full sun. Bog gardens and front of
damp borders. **Image size** One-and-a-half life-size.

This delicate creeping plant roots at the
nodes and will eventually form a dense
mat. The leaves are paired, oval and short-
stalked. The flowers are borne on relatively
long, slender stalks arising from the join of
the stem and leaf blade. They are pale pink
and cup-shaped, and 6–10mm across, but
they close when the weather becomes
cloudy. The fruits are round, splitting
around the middle to release the seeds.
Regrettably bog pimpernel is becoming
more scarce as the damp habitats that it
needs come under threat from drainage.

Common calamint [2]

Calamintha ascendens (syn. Clinopodium ascendens;
Lamiaceae or Labiatae)

Perennial. Height Up to 60cm. **Habitat** Woods,
grassland and dry banks, usually on chalk. **Distribution**
Most of northern Europe as far north as Germany. **Flowers**
July–September. **Image size** One-and-a-half life-size.

A hairy plant, with upright, square,
sometimes branched stems, calamint
spreads by means of underground stolons.
Its leaves are oval and toothed, with long
hairs flattened against the blade; they smell
of mint when crushed. The lilac to
purplish-pink flowers, 15–22mm long,
grow in a loose whorl from the leaf axils.
The petals are fused to form a two-lipped
tube, with dark spots on the lower lip. The
sepals form a five-toothed tube, the lower
two sepals being much longer than the
upper three. Calamint was used in syrups
to relieve coughs and cold symptoms. It was
also used at one time in cooking to disguise
the taint of meat that had started to go off.

Lesser skullcap [3]

Scutellaria minor (Lamiaceae or Labiatae)

Perennial. Height Up to 25cm. **Habitat** Wet heathland,
open woodland and scrub on acid soils. **Distribution**
Northern Europe as far north as southern Britain, Ireland,
Germany and southern Sweden. **Flowers** July–October.
Image size One and a half life-size.

Lesser skullcap is a delicate plant with five-
lobed flowers – pale pinkish-purple with
dark spots – borne singly or in pairs at the
junction of a leaf and the stem. The flower
projects from a tube of sepals and has a
dome-like projection on the top; the
similarity between this and the skullcaps
worn by Roman soldiers gave the plant its
common name. The genus name also refers
to this – a *scutellum* was a disc-shaped
pouch. The narrow leaves are lanceolate to
oval with a rounded or heart-shaped base;
they grow in opposite pairs, and are
occasionally toothed near the base.

Cut-leaved crane's-bill [4]

Geranium dissectum (Geraniaceae)

Straggling annual or biennial. Height Up to 50cm.
Habitat Waste and cultivated ground and grassland.
Distribution All northern Europe except Faeroe Is and
Iceland; probably not native in Scandinavia.
Flowers May–September. **In gardens** A weed.
Image size One-and-a-half life-size.

Like several other *Geranium* species, cut-
leaved crane's-bill is instantly recognisable
when you examine the leaves. These are
divided into five to seven narrow segments,
and look as though they have been through
a shredder. This doesn't seem to stop the
plant functioning as an effective weed,
managing to place itself between treasured
garden plants. The 8mm pink flowers have
a deep notch at the tip of the five petals.
Once flowering is over, the long-pointed
seedpod – the 'crane's bill' – develops. This
mechanism dries as it matures, until it
breaks away from the old flower base and
throws the seed like a catapult.

Red deadnettle [5]

Lamium purpureum (Lamiaceae or Labiatae)

Annual. Height Up to 45cm. **Habitat** Cultivated,
disturbed and waste ground, and roadsides.
Distribution All northern Europe, but only naturalised in
Iceland. **Flowers** March–October. **In gardens** A common
weed. **Image size** One-and-a-half life-size.

The leaves are enough like a stinging
nettle's (p.218) for generations of children
to torment their friends with both this and
its white cousin (p.32). Look for the square
stem, the opposite, downy, gently-toothed
leaves and the pinkish-purple, two-lipped
flowers, which are borne in whorls at the
leaf axils. Although it doesn't sting, this is a
pernicious weed. As with all successful
weeds, red deadnettle starts flowering early,
and releases copious seed to fill as many
patches as possible. In earlier times, it was
one of many treatments for 'king's evil', or
scrofula – glandular tuberculosis, thought
to be curable by the touch of a monarch.

Marsh pennywort [6]

Hydrocotyle vulgaris (Apiaceae or Umbelliferae)

Prostrate creeping perennial. Height Variable, stems to
30cm. **Habitat** Bogs, wet areas and lakesides.
Distribution All northern Europe except Iceland, Faeroe
Is and Finland. **Flowers** June–August. **Image size**
One-and-a-half life-size.

Marsh pennywort is a far-creeping, small
plant. As with bog pimpernel (see above
left), getting damp knees is essential to see
all of it. The leaves are round and fleshy,
and are carried on long stalks. Normally
plants of this family have flower heads
borne in umbrella-like inflorescences. In
pennywort this structure is reduced to a
few tiny greenish-pink flowers on a thin
stalk, which arises from the point where a
leaf-stalk joins the creeping stem. The snails
that harbour sheep liver-fluke larvae often
rest on marsh pennywort, and it has been
linked to diseased sheep and was known to
some country people as 'sheep-rot'.

Lesser snapdragon [1]

Misopates orontium (syn. *Antirrhinum orontium*; Scrophulariaceae)

Annual. Height Up to 50cm. **Habitat** Cultivated and waste land, rocky and sandy habitats and railway embankments. **Distribution** All northern Europe except Faeroe Is and Iceland. **Flowers** July–October. **Image size** Twice life-size.

An upright, delicate plant, lesser snapdragon is rare except in southern Britain, and is a weed of cultivated land. However, by germinating in spring it may be swamped in fields where a crop is sown in the autumn – an example of changes in the use of arable land affecting an annual plant. It is usually downy-hairy, with glandular hairs. The leaves are opposite or alternate, narrow or narrowly elliptical. The flowers grow in a loose spike, looking very much like small snapdragons (p.159); they have two lips and are pink or pinkish-purple. The sepal tube has lobes longer than the petals.

Dove's-foot crane's-bill [2]

Geranium molle (Geraniaceae)

Annual. Height Up to 40cm. **Habitat** Cultivated, waste and bare land, grassy habitats and dunes. **Distribution** Most of northern Europe except far north. **Flowers** April–September. **Image size** One-and-a-half life-size.

Scrutinise the rounded, lobed leaves of this crane's-bill species for long enough and they will eventually look like a dove's foot. They grow on long wayward stems, and are covered in soft downy hairs. The species name *molle* is Latin for 'soft'. This down distinguishes dove's-foot crane's-bill from the closely related shining crane's-bill (p.128), which has smooth foliage instead. In a dry autumn, the leaves turn bright red. The flowers have five pinkish-purple, deeply notched petals on slender stalks. As in other crane's-bills, they are followed by long beak-like seed heads.

Squinancywort [3]

Asperula cynanchica (Rubiaceae)

Perennial. Height Up to 40cm. **Habitat** Dry grassy places on chalk or limestone, and chalky dunes. **Distribution** Northern Europe from southern Britain and Ireland, Holland and Germany southwards. **Flowers** June–August. **Image size** Twice life-size.

The curious name comes from the use of tinctures made from the stem of the plant in a gargle for treating severe tonsillitis – or quincy (itself from the Medieval Latin *quinancia*). It is similar in general appearance to the related field madder (p.128), but is perennial and greyish-green. It has narrower leaves in whorls of four (two long and two short) up the more or less square stem. The small flowers are borne at the end of stems in sparse, loose panicles. They are four-lobed, funnel-shaped, pale lilac-pink on the outside and white on the inside.

Common dodder [4]

Cuscuta epithymum (Cuscutaceae; formerly Convolvulaceae)

Scrambling annual or short-lived perennial. Height Stems up to 60cm long. **Habitat** Parasitic, particularly on gorse, heather and clover. **Distribution** All northern Europe except Iceland and Faeroe Is. **Flowers** July–September. **Image size** Half life-size.

Dodder is parasitic on various other plants, especially gorse and heather. It used to be classified in the bindweed family, but it and a few close relatives are now placed in a family of their own. Once it germinates, the seedling will twine until it comes into contact with a stem. It then forms suckers, which penetrate the host plant and invade its structures, taking water from the roots and food from the leaves. Sometimes, stands of gorse or heather look as though they have been tied in pink string; it has no green leaves. As soon as it is well-attached, with suckers at several points, it produces dense heads of small pink, five-lobed flowers.

Upright hedge parsley [5]

Torilis japonica (Apiaceae or Umbelliferae)

Annual. Height 0.5–1.2m. **Habitat** Roadsides, hedge-banks, woodland margins, scrub and other grassy areas. **Distribution** All northern Europe except northern Scotland, Faeroe Is and Iceland. **Flowers** July–August. **Image size** One-third life-size.

One of the three common roadside members of this family, upright hedge-parsley is the last to flower, after cow parsley (p.32) and rough chervil (p.52). It is a stiff, upright plant with stiff hairs that are bent back flat against the stem. The basal leaves are long-stalked and divided into up to seven leaflets, which are also deeply cut; the upper leaves are smaller and less divided. The umbrella-shaped inflorescences are relatively narrow, with up to 12 main rays, each topped by a smaller umbel. The 2–3mm flowers are pink, white or purple-tinged, with notched petals; the outer petals are longer than the inner. Brown dry fruits follow. They are covered in hooked spines, and will travel happily hooked on to fur, feathers or clothing, ensuring that the next generation is widely distributed.

Henbit deadnettle [6]

Lamium amplexicaule (Lamiaceae or Labiatae)

Annual. Height Up to 25cm. **Habitat** Cultivated, disturbed and waste ground, normally on drier soils. **Distribution** All northern Europe except Faeroe Is; naturalised in Iceland. **Flowers** March–December. **Image size** Two-thirds life-size.

Similar in appearance to red deadnettle (see p.144), henbit deadnettle has stalkless upper leaves that clasp the stem and are fused together at the base. They are more deeply toothed. The two-lipped flowers appear in whorls at the top of the stem, and are pink to pinkish-purple. They are more slender than those of red deadnettle and the fused tube of sepals is covered in white down. Henbit deadnettle is a favoured food of chickens – hence the name.

Motherwort [1]

Leonurus cardiaca (Lamiaceae or Labiatae)

Perennial. Height 10–60cm. **Habitat** Hedge-banks, wood margins and wasteland on gravelly soils. **Distribution** All northern Europe, but naturalised and rare in Britain and Ireland. **Flowers** July–September. **Image size** One-third life-size.

The extremely hairy-backed flowers of this striking upright plant give it its rather lyrical genus name, from the Greek *leon* (lion) and *oura* (tail). Its palmate leaves, with three to seven lobes, distinguish motherwort from its many relatives, and are arranged in distinct pairs up the stems. The whorls of pink flowers can have up to 15 blooms. In the Middle Ages, the entire plant was used to treat female disorders – particularly to ease childbirth – hence the name motherwort. It was also known as a reliable remedy for fainting fits, palpitations and other cardiac conditions – hence the species name *cardiaca*.

Herb robert [2]

Geranium robertianum (Geraniaceae)

Annual or biennial. Height Up to 50cm. **Habitat** Woods, hedgerows, screes and shingle. **Distribution** All northern Europe except far north. **Flowers** May–September. **In gardens** Well-drained soil in full sun. Woodland and wild-flower gardens. **Image size** Half life-size.

In the autumn the stems and leaves of herb Robert turn scarlet. This naturally suggested to ancient herbalists that the plant could cure blood disorders, so the leaves were used to check blood flow. The redness of the leaves may be a source of the name Robert, from the Latin *ruber* (red). Other suggestions are that it was named after Robert, Duke of Normandy, after St Robert, Abbot of Molesme and founder of the Cistercian order, or after the mischievous sprite Robin Goodfellow (also known as Puck). The flowers are pink, with rounded, un-notched petals. The ferny leaves have three to five feathery lobes.

Herb Robert exudes a strong odour that is not altogether pleasant. Several folk names refer to this: stinking Bob, stink flowers and, changing gender, stinking Jenny.

Red valerian [3]

Centranthus ruber (syn. *Valeriana ruber*; Valerianaceae)

Perennial. Height Up to 1m. **Habitat** Dry sunny rocks, cliffs and old walls; sandy sites. **Distribution** Naturalised as far north as Britain, Ireland, Holland and Germany. **Flowers** June–September. **In gardens** Best in poor, dry, chalky soil in sun. Borders, dry slopes and wall crevices. **Image size** One-third life-size.

It is not easy to mistake this Mediterranean introduction, with its fleshy oval leaves and dense heads of small, tubular, five-petalled flowers. These have a long spur at the base of the tube, and are usually pink or red, but sometimes white. Red valerian will grow almost anywhere, but its parachuted seeds are expert at finding and establishing themselves in the smallest cracks. As a result, it can be seen on walls or doing its best to lever rocks apart; but it is reluctant to grow well in good soil. Butterflies and day-flying moths find the nectar irresistible – reason enough for growing it in the garden. The young leaves can be used in a salad, if the bitter taste is removed by boiling – perhaps a dish only for gourmets.

Indian or Himalayan balsam [4]

Impatiens glandulifera (Balsaminaceae)

Annual. Height Up to 2m. **Habitat** River- and canal-banks, ditches and other damp, waste and cultivated areas. **Distribution** Naturalised in most of northern Europe as far north as central Scandinavia. **Flowers** July–October. **Image size** Life-size.

Introduced to gardens from the Himalayas less than 200 years ago, this plant has become very common, thanks largely to its explosive seed capsules, which can hurl seeds more than 10m. It is welcomed in some waste areas, but a pest – albeit an attractive, sweet-scented one – that suppresses native flora in others. It is a robust plant with rather succulent stems and sharply toothed, pointed leaves in opposite pairs or whorls of three. The pale to dark purple-pink, lipped flowers, 25–40mm long, are borne in leaf axils or in loose clusters at the end of the stems. Each flower appears to have three petals, because the lower and side petal on each side are fused. There are three pink sepals; the lower one is helmet-shaped (giving the plant the nickname policeman's helmet) and has a short spur filled with nectar. This attracts bees, which are temporarily trapped in the flower, ensuring efficient pollination.

Common fumitory [5]

Fumaria officinalis agg. (Fumariaceae; sometimes included in Papaveraceae)

Scrambling annual. Height Up to 50cm. **Habitat** Waste, disturbed and cultivated ground. **Distribution** All northern Europe. **Flowers** May–October. **In gardens** A weed. **Image size** Life-size.

This successful field and garden weed is easily spotted for several reasons. The first is the blue-grey leaves, which are divided into several pairs of leaflets, each further divided into small segments. The second is the curious tubular, four-petalled flowers, 7–9mm long, which are purplish-pink with darker – almost black – tips and wings; these flowers have a blunt spur on the upper petal. The third is the round, green 2mm fruits, which contain a single seed, and the fourth is the general appearance of a large patch of the plant in one place. From afar, the bluish leaves and pink flowers look like a cloud of smoke over the ground – hence the very apt alternative name 'smoke of the earth'.

Sea rocket [1]

Cakile maritima (Brassicaceae or Cruciferae)

Annual. Height 20–45cm. **Habitat** Coastal sand, shingle and sometimes rocks. **Distribution** Northern Europe to S. Norway. **Flowers** June–August. **Image size** Life-size.

When the new volcanic island of Surtsey formed just off Iceland in 1963, botanists were eager to see which would be the first plant to appear on the totally bare new beach. It turned out to be a close, Arctic relative of sea rocket, *Cakile edentula*. This is not surprising as both plants have the ability to colonise very inhospitable areas. Sea rocket is a scruffy, branching plant with a very long taproot and succulent leaves – some linear, some pinnately lobed. Both of these features help in gathering and retaining water. The small flowers may be lilac, pink or white, and the seedpods can float on sea water without reducing the seeds' viability. It is one of a very few plants that is perfectly at home where sea and land meet, and can even survive being buried in sand.

Red bartsia [2]

Odontites verna (Scrophulariaceae)

Semi-parasitic annual. Height Up to 50cm. **Habitat** Waste places, scrub and grassy fields; also salt marshes and seashores. **Distribution** All northern Europe except Faeroe Is and Iceland. **Flowers** June–August. **Image size** Two-thirds life-size.

A much-overlooked plant, red bartsia has a dusty, careworn appearance, due to its dense covering of fine hairs. It is branched and has toothed, oblong to lanceolate leaves in opposite pairs. The small purplish-pink, two-lipped flowers are borne two at a time in the upper leaf axils. All the flowers on one stem face in the same direction. Red bartsia battens on to the roots of neighbouring grasses, taking water and minerals from them; as a result, the affected grasses' growth will be stunted.

Vervain [3]

Verbena officinalis (Verbenaceae)

Perennial. Height Up to 60cm. **Habitat** Waysides, wasteland and rough grassland, particularly on chalk. **Distribution** All northern Europe except most of Scotland, Faeroe Is, Iceland and Scandinavia. **Flowers** June–October. **Image size** Half life-size.

Vervain is not the easiest plant to spot, with its slender spikes of tiny pale pink flowers borne at the end of long branches from the square stem. The flowers are five-lobed, with two of the lobes slightly smaller than the others. The hairy leaves, in opposite pairs, are often deeply divided. The plant was clearly recognised by apothecaries, however; the genus name *Verbena* is the Latin name for a sacred bough used in sacrifices, and *officinalis* indicates its inclusion in the medical collection. It was given as a gentle diuretic, and to aid menstrual flow and lactation. An infusion was used to remedy insomnia and nervous exhaustion, to prevent stones and to give the system a general boost.

Lesser sea spurrey [4]

Spergularia marina (Caryophyllaceae)

Annual or short-lived perennial. Height Up to 30cm. **Habitat** Mud or sand in salt marshes; also salty areas inland. **Distribution** Coasts of all northern Europe except Faeroe Is. **Flowers** June–August. **Image size** One-and-a-half life-size.

Rather similar to sand spurrey (see right), this is an untidy plant, with several scrambling branches. Unlike sand spurrey, the leaves are succulent and rounded in cross-section, with a flattened top and a horny tip. The flowers grow in loose clusters, and have five pink or rose-pink petals, usually with a dark blob at the base. The five sepals are purple-tipped and usually longer than the petals.

Like its cousin greater sea spurry (*Spergularia media*), which has slightly larger, bluish-pink flowers, it is one of a select group of plants that positively like salty conditions. Both have literally taken to the roads and moved inland from the sea in some areas. This is directly related to the quantities of salt used to de-ice roads in winter. As a result, road edges have become salty turf in places, and salt-loving plants are gradually moving inland along this ribbon of suitable ground. So sea spurreys are now being seen in many inland areas.

Sand spurrey [5]

Spergularia rubra (Caryophyllaceae)

Annual or biennial. Height Up to 25cm. **Habitat** Sand and gravel, not chalky or salty. **Distribution** All northern Europe except Faeroe Is and Iceland. **Flowers** May–September. **Image size** Twice life-size.

Sand spurrey is a sprawling plant with many stems and thin pointed leaves. The flowers are in loose clusters, with the five rose-pink petals slightly shorter than the green sepals. They are visited by pollinating flies, but are always self-pollinated. Some flowers do not open at all. Until lesser and greater sea spurrey (see left) began to make their way inland, sand spurrey was the only *Spergularia* species regularly found inland. It cannot tolerate chalky or salty conditions.

Strawberry clover [1]

Trifolium fragiferum (Fabaceae or Leguminosae)

Creeping perennial. Height Stems up to 30cm long.
Habitat Grassy areas, often on clay or salty soils.
Distribution All northern Europe except Faeroe Is and
Iceland. **Flowers** June–September. **Image size**
Two-thirds life-size.

This prostrate plant looks very like white
clover (p.42) when it is not in flower,
except that the leaves do not have white
markings. The flower heads are smaller
than white clover's and are pale pink. In
fruit, the upper lip of the calyx enlarges to
form a cover that looks just like a miniature
strawberry in everything but colour –
which again is pale pink. Strawberry clover,
although reasonably common, is one of a
number of plant species showing decline
because modern farming methods have
'improved' pastureland.

Wild basil [2]

Clinopodium vulgare (Lamiaceae or Labiatae)

Perennial. Height 10–60cm. **Habitat** Dry grassy
habitats, open woodland, copse-edges, scrub and
hedgerows. **Distribution** Much of northern Europe
except far north. **Flowers** July–September.
Image size Two-thirds life-size.

Neither looking nor smelling like true
culinary basil (*Ocimum basilicum*),
although a member of the same family,
wild basil exudes a faint thyme-like
fragrance. The downy oval leaves, which are
slightly toothed, grow in pairs on upright,
downy, square stems that are typical of the
family. Bristly whorls of pink two-lipped
flowers, with a three-lobed lip, appear at
the base of the upper leaves. Wild basil
used to be carried in posies, to keep the
better-off safe from foul smells that were
deemed to carry infection. It was also
believed to repel serpents, particularly the
mythical Basilisk.

Spiny restharrow [3]

Ononis spinosa (syn. *O. campestris*; Fabaceae or
Leguminosae)

Dwarf shrub. Height 10–60cm. **Habitat** Grassy
habitats, banks, roadsides and some coastal habitats.
Distribution All northern Europe except Ireland, Faeroe Is,
Iceland and Finland. **Flowers** April–September.
Image size Two-thirds life-size.

As with other restharrows, the spiny species'
long deep roots can hinder the progress of a
field plough – the origin of the common
name. *Ononis* comes from the Greek *onos*
(ass); it is, allegedly, a favourite food of
donkeys. Spiny restharrow is more upright
and shrubby than the common species
(p.143). It has slightly greasy trifoliate
leaves, which are carried on stiff, spiny
stems. The keeled pink or purplish pea
flowers are followed by oblong seedpods.

Wild marjoram, oregano [4]

Origanum vulgare (Lamiaceae or Labiatae)

Perennial. Height Up to 60cm. **Habitat** Hedgerows,
scrub and dry grassy areas, usually on chalk.
Distribution Most of Europe except Faeroe Is, Iceland
and the far north. **Flowers** July–September. **In gardens**
Well-drained soil in full sun; spreads easily by seed. Wild
and herb gardens. **Image size** Two-thirds life-size.

On a dry, chalky bank under a hot sun, you
may smell wild marjoram before spotting it.
If you follow your nose you will find a
hairy plant with branched stems and oval
leaves. The small pink, red, purple or white
flowers crowd together in a broad head,
surrounded by purple leafy bracts. Each
flower has one notched upper lobe and
three lower. Wild marjoram has long been
used in herbal medicine, particularly in
cough mixtures. It is also, of course, a
culinary herb with an aroma strong enough
to survive drying. The more delicately
flavoured sweet and pot marjoram
(*Origanum majorana* and *O. onites*) are
closely related warm-climate plants that do
not survive in the wild in northern Europe.

Water lobelia [5]

Lobelia dortmanna (Campanulaceae)

Aquatic perennial. Height Up to 50cm above water
level. **Habitat** Stony, nutrient-poor lakes and tarns,
often in mountains. **Distribution** Britain, Ireland and
northwards from northern France. **Flowers** July–August.
Image size Two-thirds life-size.

The genus of plants named after the
Flemish botanist Mathias de l'Obel is
usually associated with the small half-hardy
annual blue- or white-flowered *Lobelia erinus*,
used in hanging baskets and for edging
borders. Water lobelia, with its single stem
and submerged rosette of blunt-tipped
linear leaves, is very different. The sparse
nodding flowers are two-lipped, pale violet
and about 15–20mm long. Water lobelia
grows only in acid water with few or no
nutrients. Lakes like this are always extremely
clear, and the plant is easy to see. It is one
of many species adapted to a very limited
habitat; in this case, it is in danger of being
lost if extra nutrients enter the water.

Corncockle [6]

Agrostemma githago (Caryophyllaceae)

Annual. Height Up to 1m. **Habitat** Cornfields and other
cultvated and waste land. **Distribution** All northern Europe.
Flowers May–August. **In gardens** Well-drained soil in full
sun. Wild-flower gardens. **Image size** Two-thirds life-size.

Sheets of dull pinkish-purple corncockle
were once so common in cornfields that
among arable farmers the plant was
considered a pest. Harvested with corn, the
seeds inevitably ended up in flour, reducing
its quality. The increased use of herbicides
and modern methods of harvesting mean
that now only a few colonies remain in East
Anglia and elsewhere in Europe. The flowers
grow on tall, wiry stems. They have five
rounded, notched petals, and narrow sepals
almost twice the length of the petals. The
petals serve as landing platforms for butterflies
and moths, which are the only insects with
long enough tongues to extract the nectar.

Alsike clover [1]

Trifolium hybridum (Fabaceae or Leguminosae)

Creeping perennial. Height Up to 40cm. **Habitat** Grassy and cultivated ground. **Distribution** Britain and Ireland, naturalised through Europe except extreme north. **Flowers** June–September. **Image size** One-and-a-third life-size.

The true origins of the Alsike clover are unknown, but this introduced species, which is thought to take its name from Alsike near Uppsala in Sweden, was cultivated in Britain as a forage crop from the 18th century. The globular flower heads are held on long upright stems and consist of a mass of florets that start purple or white and gradually turn pink, then brown. Like all clovers, the leaves consist of three oval or heart-shaped leaflets – hence the genus name *Trifolium*. They are rather like those of white clover (p.42), but have no white markings.

Cross-leaved heath [2]

Erica tetralix (Ericaceae)

Shrub. Height Up to 60cm. **Habitat** Bogs, moors, wet heathland and pine woods on acid soils. **Distribution** All northern Europe except Faeroe Is and Iceland. **Flowers** June–October. **In gardens** Moist acid soil in full sun. Ground cover and mixed borders. **Image size** Half life-size.

Cross-leaved heath likes wetter habitats than the rather similar bell heather (p.206); in fact, it's an indicator of the wetter parts of moorland or bogs. As the name suggests, the arrangement of the foliage – in whorls of four that form crosses – is also distinctive. The leaves have a shiny cuticle on the upper surface and inrolled margins. The pink (or occasionally white) flask-shaped flowers, 6–9mm long, are borne in tight clusters at the top of the stems. Like most heathers, the bush gets untidy with age. It sprawls outwards and lower branches root, forming new vigorous growth that spreads the plant farther.

Narrow-leaved everlasting pea [3]

Lathyrus sylvestris (Fabaceae or Leguminosae)

Clambering perennial. Height Stems up to 2m long. **Habitat** Woodland, scrub, embankments and hedges. **Distribution** All northern Europe except Ireland, Faeroe Is, Iceland and far north. **Flowers** June–August. **In gardens** Fertile, well-drained soil in full sun or dappled shade. Against walls, fences and hedges, and through shrubs. **Image size** Half life-size.

Each pair of this plant's distinctive long, narrow leaflets is equipped with a branched tendril. These allow it to cling to its neighbours while it scrambles on its way. Long, upright flower stalks carry racemes of rose-pink pea flowers with purplish-pink wing petals. They are similar to those of its close relative the annual garden sweet pea (*Lathyrus odoratus*), but smaller and lacking the latter's sweet scent. Narrow-leaved everlasting pea grows as a native plant in some old woods and coastal scrubland, but elsewhere wild plants are often garden escapees. They can be found making a vast improvement to usually less than attractive railway embankments.

Meadow saffron [4]

Colchicum autumnale (Liliaceae or Colchicaceae)

Perennial. Height Flowers up to 20cm; leaves up to 35cm. **Habitat** Open woods and damp meadows, often on chalk. **Distribution** Southern Britain and Ireland, France and Germany; naturalised in southern Scandinavia. **Flowers** August–October. **In gardens** Moist, well-drained soil, preferably in semi-shade. Borders, among shrubs or naturalised in grass. **Image size** Life-size.

Meadow saffron is very distinctive: it flowers long after the long, lance-shaped bright green leaves have died away. Or you could say that the flowers bloom and die away long before the leaves emerge from the earth. Whichever, the goblet-shaped flowers are 4–6cm long and have six pale pink to pale purple tepals and six orange stamens. Seeing the plant in flower makes the vernacular name 'naked ladies' seem very appropriate. However, the fact that they look rather like giant crocus flowers leads to some confusion. It is sometimes misnamed autumn crocus – a name that properly belongs to a true crocus, *Crocus nudiflorus*. The name meadow saffron is a reference to another crocus, *C. sativus*, whose three orange stamens produce the spice saffron. Meadow saffron is extremely toxic, but has been used to treat gout.

Common storksbill [5]

Erodium cicutarium (Geraniaceae)

Annual or biennial. Height Up to 50cm. **Habitat** Dry grassy places and wasteland, especially on chalk; also coastal dunes. **Distribution** All northern Europe except Faeroe Is and Iceland. **Flowers** May–September. **Image size** One-and-a-half life-size.

A very variable species, common storksbill is branched and covered in long white hairs when growing in reasonable conditions. The leaves are divided into paired leaflets, which are themselves deeply lobed. The flowers, up to 18mm across, grow in loose, long-stalked clusters; the petals are rose-pink or purplish. They are followed by a fruit with an elongated beak – the 'stork's bill' – which splits into five segments when it is ripe. Each seed falls to the ground attached to a tightly coiled remnant of the fruit, which will wind and unwind with changes in moisture. As a result, it literally screws the seed into the soil. A very short form of common storksbill, with smaller flowers, grows by coasts. This may be a subspecies, but is probably the reaction of a variable plant to a harsh environment.

Sticky catchfly [1]

Lychnis viscaria (Caryophyllaceae)

Perennial. Height Up to 60cm. **Habitat** Dry mountain cliffs and rocky places, usually on rocks of volcanic origin. **Distribution** All northern Europe except Ireland and Iceland; rare in Britain. **Flowers** May–August. **In gardens** Well-drained soil in sun. Rock gardens and scree beds. **Image size** One-quarter life-size.

This handsome plant seems to be colour-co-ordinated with its background. It is often found on cliffs of black volcanic rock, against which the narrow leaves and the spikes of bright rosy-pink 20mm flowers are shown to best effect. There is a particularly well-documented colony on the sides of Arthur's Seat – the plug of ancient volcano dominating Edinburgh. This colony has been known for over 300 years, but was gradually fading away; currently, after much care and attention, it is still holding on. The name catchfly refers to the very sticky areas on the stem at and below the leaf nodes. The finest garden cultivar is 'Splendens Plena', with bright magenta double flowers. The smaller (and smaller-flowered) alpine catchfly is sometimes listed as a variety of *Lychnis viscaria*, but is now regarded as a separate species, *L. alpina*.

Peppermint [2]

Mentha x piperata (Lamiaceae or Labiatae)

Perennial. Height Up to 90cm. **Habitat** Damp places. **Distribution** Naturalised throughout northern Europe. **Flowers** July–October. **In gardens** Any well-drained moist soil in sun or partial shade. Best grown in a container. **Image size** One-third life-size.

A hybrid between water mint (p.202) and spearmint (right), peppermint is very variable as a result. It may be hairless or hairy, and is often purple-tinged. The leaves may be varied shapes, from oblong to oval, and are sharply cut. The flower heads are in whorls, around the bases of the upper leaves. The small, lilac, two-lipped flowers are sterile. The commonest variety smells strongly of peppermint, but there is also a hairless variety whose scent is reminiscent of eau de Cologne.

Spearmint [3]

Mentha spicata (Lamiaceae or Labiatae)

Perennial. Height Up to 90cm. **Habitat** Waste ground, roadsides and damp places. **Distribution** Naturalised in most of northern Europe except Iceland and far north. **Flowers** August–September. **In gardens** Any well-drained moist soil in sun or partial shade. Best grown in a container. **Image size** Life-size.

Many gardeners who grow no other herb have a patch or pot of this mint, the perfect complement to roast lamb. It is a very variable species, particularly in the degree of hairiness. The paired leaves vary from lanceolate to oval, sometimes broadly so, and from deep green and shiny to greyish, wrinkled and downy. The small lilac flowers, which are borne in the axils of leaf-like bracts, form a long spike-like head. As you might expect from such a promiscuous group, spearmint is of hybrid origin, arising from a cross between round-leaved mint (*Mentha suaveolens*) and horse mint (*M. longifolia*). It is fertile, so there must have been a doubling of the chromosome number at some stage, enabling the production of fertile seed. It is very vigorous in gardens – from where it escaped into the wild – so confine its roots in a sunken or free-standing container if you don't want it to take over.

Marsh mallow [4]

Althaea officinalis (Malvaceae)

Perennial. Height Up to 1.3m. **Habitat** Salt-marsh edges, ditch-sides, grassy places and banks, usually near the sea. **Distribution** Britain (except Scotland), Ireland, Denmark and Germany southwards. **Flowers** July–September. **Image size** One-sixth life-size.

The name is no coincidence – the sweet known as marshmallow was once made from this plant's thick, fleshy rootstock, which is rich in gummy mucilage. Marsh mallow is a tall, densely hairy plant with rounded lower leaves on a long stalk. The upper leaves are stalkless, lobed and more or less triangular in outline. The pale lilac-pink flowers are borne in groups of up to three arising from a leaf axil, and are up to 5cm across, with five broad petals. The roots have been used as a specific remedy for inflamed gums and sore throats.

Cheddar pink [5]

Dianthus gratianopolitanus (syn. *Dianthus caesius*; Caryophyllaceae)

Creeping perennial. Height 10–25cm. **Habitat** Exposed cliffs, rocks and screes, especially on limestone. **Distribution** Northern Europe as far north as south-west England (rare), Belgium and central Germany. Not found in Ireland. **Flowers** May–July. **In gardens** Well-drained alkaline soil in full sun. Rock gardens, raised beds and alpine troughs. **Image size** Twice life-size.

This densely tufted plant has adapted perfectly to its less than friendly habitat of limestone rocks baked in sunshine. It has linear grey-green leaves, which prevent water loss, and long creeping sterile shoots which form a dense mat. Erect flowering stems carry pink flowers up to 30mm across, which are sweetly clove-scented. Cheddar pink is a protected species, listed in the Red List under Britain's Wildlife and Countryside Act and, as a result, seems to be surviving in its few British localities, including Cheddar Gorge, from which it is named. It almost certainly survived because the precipitous limestone cliffs where it grows would have been free of trees when most of the country was under wild wood. Surprisingly, pinks are not named for their colour but for the ragged – or 'pinked' – appearance of their petals.

Snapdragon [1]

Antirrhinum majus (Scrophulariaceae)

Annual or short-lived perennial. Height Up to 80cm. **Habitat** Rocky and waste ground, railway embankments and walls. **Distribution** Naturalised in Britain and Ireland, Holland and Germany southwards. **Flowers** June–September. **In gardens** Most soils in sun or part-shade. Beds and borders. **Image size** One-and-a-half life-size.

In northern Europe, snapdragons grow only as escapees from cultivation. They are well-known garden plants with narrowly oval, untoothed leaves and multi-coloured flowers whose two lips form a 'mouth' that pollinating bees must push open for access. Children, of course, simply squeeze the sides of the flowers together.

Snapdraons are normally treated in gardens as annuals, but they produce plenty of seed from the conical fruits and occasionally jump the fence to take up life as a short-lived perennial. They are particularly happy growing on walls, where they are in similar conditions to the warm rocky slopes of their native southern Europe. Escaped plants will revert to producing pink to pinkish-purple, or occasionally yellowish-white, flowers.

Knotgrass [2]

Polygonum aviculare (Polygonaceae)

Scrambling annual. Height Stems up to 2m long. **Habitat** Bare, waste and cultivated land; seashores. **Distribution** All northern Europe. **Flowers** June–October. **In gardens** A weed. **Image size** Half life-size.

This is a perfect example of the apt use of the much-misused term 'weed'. Knotgrass, with oval leaves up to 50mm long and the tiniest of flowers, sprawls over any disturbed ground. Looking at it closely through a lens reveals a more interesting plant than first seems likely. The junction between leaf and stem is covered by a papery sheath, and these give the plant a jointed appearance – the genus name *Polygonum* means 'many joints'. The pink,

whitish or greenish flowers grow in clusters of up to six in the axils of upper leaves. Each consist of five perianth segments and produces a dull brown triangular fruit. These are much loved by seed-eating birds, which are often be seen feeding on knotgrass. The species name *aviculare* is from the Latin for 'small bird'.

Common valerian [3]

Valeriana officinalis (Valerianaceae)

Perennial. Height 1–2m. **Habitat** Rough grassland, usually damp but sometimes on dry chalk. **Distribution** All northern Europe. **Flowers** June–August. **In gardens** Moist, well-drained soil in sun or semi-shade. Back of borders. **Image size** One-third life-size.

This plant is easily distinguished from marsh valerian (p.127) by its size and pinnate leaves, the lower ones on long stalks; its leaflets are toothed. The pink or whitish flowers grow in dense, broad panicles or rounded clusters, at the top of the stems. The flowers are five-petalled, with a pouch at the base of the flower tube; unlike those of marsh valerian, they are bisexual. Valerian has been much used herbally, particularly as a general nerve tonic or to cure anxiety and to relieve the symptoms of St Vitus' dance and epilepsy. The roots have a very strong smell, reminiscent of new leather and are attractive to cats. Dried roots were placed in linen drawers.

Water avens [4]

Geum rivale (Rosaceae)

Semi-aquatic perennial. Height Up to 60cm. **Habitat** Marshy ground, streamsides, damp woodland and meadows, and wet rocks. **Distribution** All northern Europe. **Flowers** April–September. **In gardens** Moist soil in sun or shade. Front of borders. **Image size** Half life-size.

A softly hairy plant, water avens has basal leaves divided into as many as six pairs of

different-sized, toothed leaflets. The stem leaves are usually three-lobed. The rather sparse, nodding bell-shaped flowers are borne at the end of the stems. They have creamy-pink to pinkish-purple petals, and purple sepals. They are followed by fruits with hooked styles. Water avens hybridises with herb bennet (wood avens; p.99) where the two grow within range of each other. The resulting hybrids are fertile and may form complex hybrid populations.

English stonecrop [5]

Sedum anglicum (Crassulaceae)

Perennial. Height Up to 10cm. **Habitat** Rocks, shingle, dunes and short grass, usually not on chalk. **Distribution** Western parts of northern Europe except Germany and Iceland. **Flowers** May–August. **In gardens** Sharply drained, rather poor soil. Rock gardens and alpine troughs. **Image size** Life-size.

This small mat-forming plant is ideally adapted to living in conditions where water is hard to come by – either because the drainage is so good that the roots have difficulty in extracting water, or because of salty conditions. In all cases, English stonecrop is a master of water-retention. Its small succulent leaves are covered with a grey-green waxy cuticle to prevent drying, and the whole plant grows flat to the ground to prevent wind-scorch. Even the small, pink-flushed white flowers are waxy. Occasionally the leaves will turn red, indicating that the plant is growing in a particularly sunny spot. The red pigment is a means of protecting its green chlorophyll, which can be broken down by excess sunlight.

Long-headed poppy [1]

Papaver dubium (Papaveraceae)

Annual. Height Up to 60cm. **Habitat** Arable and waste land, and roadsides. **Distribution** All northern Europe except Iceland and Norway. **Flowers** June–August. **In gardens** Fertile, well-drained soil in full sun. Annual, mixed and herbaceous borders, and wild-flower gardens. **Image size** One-fifth life-size.

Distinctive elongated seedpods, like tiny pepper pots, distinguish this species from other cornfield poppies, such as the common or field poppy (see right), and give it its common name. It also has smaller, paler red flowers, usually without a dark centre, and violet (rather than blue-black) anthers. On close inspection, the stem hairs lie against the slender flower stems rather than sticking out at right angles, as in the field poppy. The grey-green foliage is pinnate and lobed and arranged alternately up the stem. The flowers appear all summer long, so it more than earns its keep as a garden plant.

Red clover [2]

Trifolium pratense (Fabaceae or Leguminosae)

Perennial. Height Up to 60cm. **Habitat** Grassy habitats, waste and cultivated land. **Distribution** All northern Europe except parts of far north; often naturalised. **Flowers** May–September. **In gardens** Moist, well-drained soil in full sun. Wild-flower gardens. **Image size** One-fifth life-size.

Despite the name, red clover has flowers in rounded heads that err on the side of pink or reddish-purple rather than true red. The foliage is distinguished by its narrow, pointed leaflets, which are marked with a pale 'V'. Red clover is an important plant, not least because it fixes nitrogen from the air via nodules on its roots and turns it into salts, which plants can then absorb. As a result, red clover has long been widely grown as a field crop, and from there it has spread far beyond its native habitats. It is also a vital source of nectar for bumblebees, and in some parts of Britain goes by the name of 'bee bread'.

Orange balsam [3]

Impatiens capensis (Balsaminaceae)

Annual. Height Up to 1.5m. **Habitat** River and canal banks. **Distribution** Naturalised in parts of Britain, Ireland and France. **Flowers** June–August. **Image size** Half life-size.

The tangerine-coloured flowers of orange balsam look out of place under leaden northern European skies, suggesting exotic origins. Like all the balsams found in Britain, except the yellow-flowered touch-me-not balsam (*Impatiens noli-tangere*), it is in fact an introduction, originally from North America. It is almost certainly a garden escapee, although it is little grown today. The flowers have brownish freckles and a distinctive hook-like spur. The oval leaves are arranged alternately rather than in the whorls of Indian balsam (p.148). Orange balsam's long, narrow seed capsule explodes to the touch, catapulting its seed several metres.

Small-flowered catchfly [4]

Silene gallica (Caryophyllaceae)

Annual. Height Up to 45cm. **Habitat** Waste and cultivated land on sandy soil. **Distribution** Parts of Britain and Ireland, France, Belgium, Holland and Germany. **Flowers** June–October. **Image size** Life-size.

Several catchfly species have sticky hairs, which are designed both to snare pollinating insects – the origin of the name catchfly – and to guarantee that the insect moves on to the next plant with pollen glued to its body. Small-flowered catchfly produces tiny pinkish-red or white flowers typical of the genus, with five notched petals and a distinctive sticky-hairy calyx, but on a miniature scale – they are only 6–10mm across. Slightly nodding, the flowers are held in a one-sided, forked spike. The narrow lanceolate leaves grow in opposite pairs. Small-flowered catchfly is very rare, a casualty of the over-use of nitrogenous fertilisers on arable land; it should not be picked or dug up.

Water figwort [5]

Scrophularia auriculata (Scrophulariaceae)

Perennial. Height Up to 1.2m. **Habitat** River, lake and stream margins, marshes, fens and other wet or moist places. **Distribution** Britain, Ireland, Belgium, Holland, France and southern Germany. **Flowers** June–September. **Image size** Life-size.

Water figwort bears brown-red flowers on square, reddish stems, similar to those of common figwort (p.221). However, it is taller than its relative, and close inspection reveals wings on the angles of the stems. Its foliage looks rather like that of betony (p.206), so at one time it was called water betony. The leaves are more rounded than common figwort's, with rounded teeth and two small lobes at their base. Like the common species, water figwort has root nodules and was used to treat scrofula.

Common, field or corn poppy [6]

Papaver rhoeas (Papaveraceae)

Annual. Height Up to 60cm. **Habitat** Arable and waste land and roadsides. **Distribution** All northern Europe except far north. **Flowers** June–October. **In gardens** Fertile, well-drained soil in full sun. Wild-flower gardens and borders. **Image size** Life-size.

The common poppy is the cherished symbol of remembrance for those whoe fought and died in the First and Second World Wars. Yet in ancient times it was also associated with new life, particularly in farming. Poppy seeds have been found in Egyptian grain stores from 2500BC and the Romans held the poppy sacred to the corn goddess Ceres. In both France and Britain there was a belief that picking the flowers would induce a thunderstorm. Held aloft on upright, hairy stems, the solitary red flowers, up to 10cm across, have papery, overlapping petals with a black blotch at their base. They last for only one day and are followed by rounded seed capsules with pores below the flattish cap. The roughly hairy leaves are deeply cut with narrow lobes.

Broad-leaved everlasting pea [1]

Lathyrus latifolius (Fabaceae or Leguminosae)

Climbing or scrambling perennial. Height Stems up to 3m long. **Habitat** Grassy places, banks, hedgerows, scrub and roadsides. **Distribution** France. Naturalised in Britain, Belgium, Holland and Germany. **Flowers** July–September. **In gardens** Well-drained soil in full sun or dappled shade. A vigorous climber for fences, walls or through shrubs. **Image size** One-and-one-third life-size.

The everlasting pea is a French native that has been grown in gardens elsewhere since at least the 15th century. It is still better known as a garden plant, and wild populations probably escaped from there. It climbs on vigorous winged stems. The leaves are blue-green and consist of a single pair of deeply veined, linear or elliptical leaflets; the rest are reduced to tendrils, by which it clings to its supports. The long-stalked pea flowers, up to 3cm across, are normally bright magenta-pink, although shades of purple and pink through to white are found. They are followed by long brown seedpods. White-flowered garden cultivars include 'White Pearl' and 'Blushing Bride' – the latter's flowers flushed with pink.

Red-veined or wood dock [2]

Rumex sanguineus (Polygonaceae)

Perennial. Height Up to 60cm. **Habitat** Damp, usually shady places such as woods; hedgerows. **Distribution** All northern Europe. **Flowers** June–September. **Image size** One-and-one-third life-size.

Docks are notoriously difficult to tell apart, and red-veined or wood dock is no exception. It bears a strong resemblance to clustered dock (p.225). The main differences lie in the stem, which is green rather than reddish and more upright rather than zig-zagged. The leaves, however, are similar: oblong with a rounded base and almost always with green veins – despite the common name. The flowers are held in pendulous clusters, and are tiny, insignificant and reddish-green in colour.

Common sea lavender [3]

Limonium vulgare (Plumbaginaceae)

Perennial. Height Up to 40cm. **Habitat** Muddy salt marshes. **Distribution** All northern Europe as far north as south-western Sweden. **Flowers** July–September. **Image size** One-and-one-third life-size.

Greyish-green and leathery, the foliage of common sea lavender enables it to cope with its salt marsh habitat. In summer, it weaves a stunning reddish, lavender-lilac tapestry along with its companions the rock sea lavender (p.190) and low-growing mats of sea lavender (*Limonium bellidifolium*). Sea lavender's tiny flowers appear on branched stems as one-sided spikes and attract pollinating beetles, bees and flies.

Annual seablite [4]

Suaeda maritima (Chenopodiaceae)

Annual. Height Up to 50cm. **Habitat** Salt marshes. **Distribution** All northern Europe except northern Scandinavia. **Flowers** July–September. **Image size** One-and-one-third life-size.

Annual seablite is a halophyte – adapted to tolerate high concentrations of salt in water. It flourishes in muddy salt marshes where it is completely immersed in seawater twice a day. Consequently, it has evolved long, narrow, succulent, water-conserving leaves on tough but reddish flexible upright stems, designed to endure the tidal flow. The tiny green flowers emerge at the leaf axils and are followed by a reddish fruit.

Marsh lousewort, red rattle [5]

Pedicularis palustris (Scrophulariaceae)

Semi-parasitic biennial or annual. Height Up to 60cm. **Habitat** Wet heaths, bogs and moors. **Distribution** All northern Europe except Iceland. **Flowers** May–September. **Image size** One-and-one-third life-size.

Both this species and common lousewort (p.136) were blamed for infecting cattle with fluke worm and sheep with lice; *Pedicularis* comes from the Latin *pedis* (louse). Red rattle refers to the rattling sound made by the seeds in their capsule. A semi- (or hemi-) parasitic, this species produces food by photosynthesis but also uses its roots to draw vital nutrients from grasses. The upright reddish-brown stems carry reddish-pink flowers, each with two lips; the upper forms a prominent hood and the lower has three lobes. The flowers and fruits have distinctive inflated calyces, which are covered in fine hairs. The foliage is fern-like and consists of opposite pairs of lobed leaflets. Marsh lousewort is declining in Britain due to the drainage of its habitats.

Montbretia [6]

Crocosmia x *crocosmiiflora* (garden syn. *Montbretia crocosmiiflora*; Iridaceae)

Spreading cormous perennial. Height 50–60cm. **Habitat** Woodland, hedgerows, wasteland and cliffs. **Distribution** Widely naturalised in Britain, Ireland and France, especially in mild-climate areas. **Flowers** July–September. **In gardens** Moist, well-drained soil in full sun or partial shade. Mixed borders and between shrubs. **Image size** Life-size.

Montbretia is a hybrid between *Crocosmia aurea* and *C. pottsii*, both South African natives, that was introduced to Europe as a garden plant. The orange-red or occasionally yellow flowers have a long, narrow tube, six oval, spreading petals and protruding stamens. Much of their beauty lies in the way the flowers are held on arching two-sided spikes with tightly furled buds and open blooms together on one spike.

Crimson clover [1]

Trifolium incarnatum (Fabaceae or Leguminosae)

Annual. Height Up to 50cm. **Habitat** Fields and
wasteland. **Distribution** All northern Europe except
Ireland, Faeroe Is and Iceland. **Flowers** May–September.
In gardens Moist, well-drained soil in full sun.
Wild-flower gardens. **Image size** Half life-size.

There is no mistaking crimson clover's
astonishing blood-red flowers. Aside from
their vibrant colouring, the flower heads are
distinct from those of other clovers, being
more oblong than rounded. The trifoliate
leaves consist of three oval to rounded
leaflets with tiny teeth at their tips.
Crimson clover was once cultivated as a
fodder crop and is thought to originate
from southern Europe, although it was
recorded as a garden plant in Britain as
long ago as the 16th century.

Fox and cubs, orange hawkweed [2]

Pilosella aurantiaca (syn. *Hieracium aurantiacum*;
Asteraceae or Compositae)

Perennial. Height Up to 50cm. **Habitat** Grassland,
meadows, and waste and cultivated land.
Distribution Native to France, Germany and Scandinavia.
Widely naturalised elsewhere in northern Europe.
Flowers June–August. **In gardens** Poor or moderately
fertile, well-drained soil in full sun or partial shade.
Wild-flower gardens and meadows, walls and banks.
Image size Half life-size.

Fox and cubs gets this charming name from
the clusters of luminous 15mm orange-red
or orange-brown flowers and their buds,
which grow together on tall, slender stems.
They are also the reason why this plant is
widely cultivated in gardens – from where
it has escaped in many areas. The
dandelion-like flowers consist of ray florets
with toothed tips. The buds are covered in
black hairs, as are the leaves and stems.
The foliage is bluish-green, elliptical or
lance-shaped and mainly arranged in a
rosette at the base of the stem. The plant
spreads on leafy runners.

Marsh cinquefoil [3]

Potentilla palustris (syn. *Comarum palustre*; Rosaceae)

Creeping perennial. Height 50cm. **Habitat** Bogs,
marshes, fens, ditches and wet meadows.
Distribution Throughout northern Europe.
Flowers May–July. **Image size** Two-thirds life-size.

Marsh cinquefoil is unique among
cinquefoils for its maroon flowers – most
of its relatives produce yellow flowers, the
other exception being the white-flowered
rock cinquefoil (p.47). The flowers consist
of five pointed petals and rather extravagant
reddish sepals, which are significantly larger
than the petals and have a light covering of
hairs on the outside. The stems too are
reddish, upright and branched. The leaves
consist of two or three pairs of leaflets with
toothed edges – *cinquefoil* is derived from
the French meaning 'five leaves'. Marsh
cinquefoil spreads on creeping underground
stems or rhizomes, making a tangled mass
of stems, leaves and foliage wherever it goes.

Salad burnet [4]

Sanguisorba minor (Rosaceae)

Perennial. Height Up to 80cm. **Habitat** Dry grassland,
rocky places and banks. **Distribution** All northern Europe
except far north. **Flowers** May–September.
In gardens Well-drained soil in full sun or light shade.
Herb and vegetable gardens, containers and herbaceous
borders. **Image size** Half life-size.

The evergreen, rounded and toothed
leaflets of salad burnet are arranged in pairs
on chunky grooved stems. The flower heads
are rounded too – a feature that
distinguishes this species from great burnet
(p.217). Male, female and bisexual flowers
are all crammed on to one 15mm flower
head – the yellowish male or bisexual
flowers at the lower end, the reddish
females at the upper. The young leaflets
taste faintly of cucumber, and can be added
to winter salads, cold drinks, casseroles and
soups. If you plant salad burnet in a
container you can harvest the fresh young

leaves all year long, but it is also handsome
enough to make a striking presence in a
perennial border.

Scarlet pimpernel [5]

Anagallis arvensis (Primulaceae)

Annual. Height Up to 20cm. **Habitat** Cultivated, waste
and disturbed land, and sandy habitats.
Distribution All northern Europe except far north.
Flowers May–October. **Image size** Life-size.

As regular as clockwork on bright days, the
perky flowers of scarlet pimpernel open at
eight in the morning and close at three in
the afternoon. If, however, the weather is
cloudy or wet, they are liable to shut up at
any moment – hence the country names
old man's weathervane and poor man's
weatherglass. It is a plant of square
sprawling stems and solitary flowers.
These consist of five rounded, overlapping
petals, which are usually red but sometimes
pink, lilac or blue; they maintain a
complete ring when they drop. The stems
droop as the globular fruits form. Opposite
pairs of shiny, oval leaves grow in pairs
along the stems; they have tiny black dots
beneath. It was once believed that scarlet
pimpernel could cure madness, relieve
misery and bring laughter.

Common lungwort [1]
Pulmonaria officinalis (Boraginaceae)

Perennial. Height Up to 30cm. **Habitat** Woods, hedge-banks and other shady places. **Distribution** As far north as southern Sweden; introduced in Britain. **Flowers** March–May. **In gardens** Moist soil in full sun or partial shade. Shady borders, and woodland and wild-flower gardens. **Image size** One-third life-size.

White-spotted and supposedly looking like diseased lungs, lungwort's oval leaves were used to treat all manner of lung ailments. In fact the leaves contain silicic acid, which has been shown to restore elasticity of the lungs. There was a belief that the white spots were drops of the Virgin Mary's milk or tears, and several folk names refer to this, e.g. Virgin Mary's milkdrops The flowers, arranged in loose cymes, start as pink buds, then turn reddish and later blue as they open and mature. They are funnel-shaped, with five lobes and a long hairy calyx.

Lesser periwinkle [2]
Vinca minor (Apocynaceae)

Creeping sub-shrub. Height Stems up to 1m long. **Habitat** Woods and shady banks. **Distribution** Northern Europe as far north as Denmark. **Flowers** July–September. **In gardens** Any well-drained soil in sun or partial shade. Ground cover in woodland gardens and shrub borders. **Image size** One-third life-size.

A ubiquitous plant of country lanes, lesser periwinkle is probably an introduction from the Continent rather than a true British native. The long, sprawling stems weave together to form thick mats of foliage, making it an invaluable ground-cover plant in gardens. The evergreen leaves are oval, smooth and leathery, and are arranged in opposite pairs. Long flower stems arise from the leaf nodes and poke their way through the foliage, carrying solitary blue-violet flowers at the tips, each with five uneven-sized petals and a white central eye. Lesser periwinkle roots at each leaf node, unlike greater periwinkle (see above right).

Greater periwinkle [3]
Vinca major (Apocynaceae)

Spreading sub-shrub. Height Stems up to 1.5m long. **Habitat** Hedgerows, scrub and woodland. **Distribution** Naturalised in Britain, Ireland and France to Denmark. **Flowers** March–May. **In gardens** Dryish soil in full sun. Ground cover, woodland gardens, shady banks and borders. **Image size** Life-size.

The greater periwinkle was introduced in northern Europe from the Mediterranean, and escaped from gardens to colonise hedgerows and woodland. Spreading, arching stems, which root at the tips, produce pairs of glossy, bright green, oval leaves. The flower stems emerge at the leaf axils, each bearing a solitary 3–5cm flower of five violet-blue petals. Periwinkles were thought to have many medical uses – in France it was called *toute-saine* (all-heal). It makes good ground cover in the garden, and there are many attractive cultivars, some with variegated foliage.

Grape hyacinth [4]
Muscari neglectum (syns. *M. atlanticum*, *M. racemosum*; Liliaceae)

Bulbous perennial. Height Up to 30cm. **Habitat** Dry grassland, hedge-banks and cultivated land. **Distribution** Britain (rare), France and Germany. **Flowers** March–May. **In gardens** Well-drained soil in full sun. Wild-flower and deciduous woodland gardens, shrubberies and lawns. **Image size** Life-size.

The common grape hyacinth once flourished on dry heathland in Britain, but it has been sacrificed for housing and other developments and is probably now only native in East Anglia. Yet it flourishes in gardens, where it spreads at great speed to form carpets of blue. The flowers, arranged in a spike-like raceme, are blackish-blue to deep violet. Each is a grape-like tube up to 7mm long with a row of white teeth at the tip. The grass-like leaves are channelled, so they too almost form a tube, and have a reddish tinge at their base.

Snake's-head lily or fritillary [5]
Fritillaria meleagris (Liliaceae)

Bulbous perennial. Height Up to 50cm. **Habitat** Damp grassland. **Distribution** Britain, France, Germany and Holland. **Flowers** April–May. **In gardens** Moist soil in full sun or light shade. Damp borders and woodland gardens. **Image size** One-third life-size.

The deep plum-coloured flowers of the snake's-head fritillary could be fashioned from the softest suede. Printed with a distinctive chequerboard pattern – which gives it one of its oldest folk names, dice-box – you need to get down on hands and knees and look up its skirt for the best effect, as the nodding flowers are like miniature lanterns that hang on upright slender stems. Narrow lanceolate leaves clasp these stems. It's a fallacy that these beautiful plants are difficult to grow in gardens. So long as they are grown among grasses or other plants that shade their roots in summer, they will thrive. It's worth the risk.

Sweet violet [6]
Viola odorata (Violaceae)

Creeping perennial. Height Up to 15cm. **Habitat** Woods, hedgerows and scrub. **Distribution** All northern Europe except far north. **Flowers** February–May. **In gardens** Moist but well-drained fertile soil in full sun or partial shade. Woodland gardens and front of borders. **Image size** Two-thirds life-size.

Sweet violet is one of the first wild flowers to bloom in the wake of winter, and never fails to lift the spirits – especially as it exudes a delicious scent. The deep purple (or sometimes white) flowers have five petals, with the lower petal forming a lip and extending behind into a short spur. The heart-shaped evergreen leaves grow in a rosette on long, slender stems. To the ancient Greeks it was the symbol of Athens and also the flower of Aphrodite, the goddess of love. Later, the petals were strewn on cottage floors as an air-freshener. It is still used in toiletries and confectionery.

Green alkanet [1]

Pentaglottis sempervirens (syn. *Anchusa sempervirens*; Boraginaceae*)*

Perennial. Height Up to 1m. **Habitat** Meadows, wasteland, woodland and hedgerows. **Distribution** Mostly naturalised in northern Europe as far north as southern Norway and Sweden. **Flowers** May–September. **In gardens** Damp soil in full or partial shade. Wild-flower and woodland gardens. **Image size** Half life-size.

Green alkanet is generally thought to have been introduced to northern Europe from France or Spain in the Middle Ages. The name originates from the Arabic *al-henna*, and it was once cultivated for the henna-like red dye that can be extracted from the roots. The stems are upright and covered in bristles, as are the alternate oval leaves. The pretty funnel-shaped flowers are bright blue with a startling white eye; they appear on coiled cymes. Green alkanet is reminiscent of forget-me-nots, and as such is prized by gardeners; but it tends to be invasive if not kept under control.

Ivy-leaved speedwell [2]

Veronica hederifolia (Scrophulariaceae)

Sprawling annual. Height Up to 60cm. **Habitat** Cultivated land. **Distribution** All northern Europe except Faeroe Is and Iceland. **Flowers** March–August. **Image size** Two-thirds life-size.

There are around 200 species of speedwell, but this is the one that most people are likely to recognise, being common throughout most of northern Europe. Ivy-leaved speedwell lives up to its name, with pale green kidney-shaped and lobed foliage that resembles ivy leaves. Both stems and leaves are covered in hairs. The tiny solitary, pale blue flowers are held on long, slender flower stalks from the leaf axils. The subspecies *lucorum* has flowers that are purple-lilac to white.

Scottish primrose [3]

Primula scotica (Primulaceae)

Perennial. Height 5–6cm. **Habitat** Grassy heaths, coastal grassland and dunes. **Distribution** Northern Scotland and Orkney Is. **Flowers** May–June. **Image size** Two-thirds life-size.

The rare Scottish primrose is unusual among primroses in having flowers that are all the same, rather than having pin-eyed and thrum-eyed forms as in the primrose (p.84). As a result, each flower may be either cross- or self-pollinated. Deep purple, they have a bright yellow eye at their centre and are held in clusters on upright stems. The leaves are arranged in a rosette at the base of the flower stem, and both leaves and stem are covered in a white, mealy bloom. The scarcity of this flower is attributed to the very particular growing conditions it needs, and these are found only in the north of Scotland and the Orkneys.

Water forget-me-not [4]

Myosotis scorpioides (Boraginaceae)

Spreading perennial. Height Up to 70cm. **Habitat** Pond edges and damp grassland. **Distribution** All northern Europe. **Flowers** May–September. **In gardens** Wet soil or shallow water to 10cm deep in full sun or partial shade. Pond margins and bog gardens. **Image size** Life-size.

Forget-me-nots used to be called scorpion grass – the flowers are borne in cymes that curl over like a scorpion's tail as the buds open. (*Scorpioides* means 'scorpion-like'.) According to German folklore, a knight fell into a river while picking flowers for his lover; he cried '*Vergisz mein nicht*' ('Forget me not') as he drowned, and the plant came to symbolise faithfulness in love. *Myosotis* ('mouse-ear') refers to the pointed foliage tips. Each flower of this species has five notched sky-blue (or sometimes pink or white) petals and an eye which can be white, pink or yellow. The slender, angular stems are slightly hairy, with slim, smooth leaves alternately up their length.

Germander speedwell [5]

Veronica chamaedrys (Scrophulariaceae)

Spreading perennial. Height Up to 50cm. **Habitat** Grassland, woodland, hedgerows and wasteland. **Distribution** All northern Europe. **Flowers** March–July. **Image size** Life-size.

This, the most common species of speedwell, is conspicuous for its sprightly four-petalled blue flowers, 10mm across, each with a white eye, on sprawling, hairy stems. The leaves are dark green and deeply toothed, and arranged in opposite pairs. Two rows of long hairs up the stem help to keep unwelcome insects at arm's length. The structure of the flower, however, is adapted for pollination by flies, which grasp the stamens between their legs and are sprung forward into a position where their abdomen is dusted with pollen while they help themselves to nectar. Speedwell tends to colonise roadsides, so was thought to speed travellers on their journey; otherwise, the name alludes to its curative properties for diseases such as smallpox and measles.

Wood forget-me-not [6]

Myosotis sylvatica (Boraginaceae)

Biennial or perennial. Height Up to 50cm. **Habitat** Woodland and mountain grassland. **Distribution** All northern Europe as far north as southern Scandinavia. **Flowers** April–July. **In gardens** Well-drained soil in full sun or partial shade. Mixed borders or wild-flower gardens. **Image size** Life-size.

This species has lance-shaped leaves, covered in soft hairs and arranged in a rosette at the base of lanky flower stems. The flowers are typical of forget-me-nots, with five rounded petals of a vibrant sky-blue and a deep yellow eye, but are unusually large at up to 10mm across. The wood forget-me-not has been the source of many perennial garden cultivars, including 'Snowball', with white flowers; 'Blue Basket', with upright stems of azure blooms; and the compact rose-pink 'Pompadour'.

It is illegal and unnecessary to dig up bluebells from the wild. Control vigorous populations in the garden by removing faded flower heads to prevent self-seeding.

Common dog violet [2]

Viola riviniana (Violaceae)

Perennial. Height Up to 20cm. **Habitat** Open woodland, grassland and banks. **Distribution** All northern Europe. **Flowers** April–June. **In gardens** Moist, well-drained soil in full or partial shade. Woodland and wild-flower gardens. **Image size** Two-thirds life-size.

Dog violets are unscented, suggesting that they are inferior to scented species such as the sweet violet (p.167). The common dog violet makes delightful straggly clumps of 15–25mm violet-blue flowers, each with five petals and a pale purple or whitish spur. The semi-evergreen heart-shaped leaves grow, like the flowers, on trailing, slender stems. As with most violets, the early flowers set only a little seed; small flowers formed later in the season do not open and are completely self-pollinated. Despite its invasive habit, the stunning dark purplish-green leaves of *V. riviniana* 'Purpurea Group' make it essential for a woodland garden.

Ground ivy [3]

Glechoma hederacea (Lamiaceae or Labiatae)

Creeping perennial. Height Up to 15cm. **Habitat** Hedgerows, woods, grassland and wasteland. **Distribution** All northern Europe. **Flowers** March–May. **In gardens** Moist, well-drained soil. Hanging baskets and containers. **Image size** Two-thirds life-size.

Clusters of pale violet flowers with a spotted lower lip hint at the ground ivy's close relationship with catmint (p.56). The flowers and paired kidney-shaped, toothed leaves are arranged along hairy creeping – and usually invasive – stems. The whole plant has a strong balsamic odour. For centuries ground ivy has been used to cure coughs and headaches and, in the form of a herbal tea, as a digestive tonic.

English bluebell [1]

Hyacinthoides non-scripta (syns. *Endymion non-scriptus, Scilla non-scripta*; Liliaceae)

Bulbous perennial. Height Up to 40cm. **Habitat** Woods, heathland and scrub. **Distribution** Native to Britain, Ireland and western France. Naturalised in other parts of northern Europe. **Flowers** April–June. **In gardens** Moist, well-drained soil in partial shade. Wild and woodland gardens. **Image size** Two-thirds life-size.

In late spring deciduous woodlands and copses erupt in a haze of blue as the English bluebell comes into flower – a sight the poet Alfred Tennyson described as 'the heavens upbreaking through the earth.' It produces glossy, long linear leaves and violet-blue, tubular pendulous flowers carried in a one-sided raceme. These distinguish it from the tenacious and invasive Spanish bluebell (p. 181), which has bell-shaped flowers on a two-sided raceme. A supreme competitor, the Spanish bluebell and its hybrids currently pose a serious threat to the English species.

Ivy-leaved toadflax [4]

Cymbalaria muralis (Scrophulariaceae)

Trailing perennial. Height Stems up to 60cm long.
Habitat Walls, rocks and woodland. **Distribution**
Naturalised in northern Europe as far north as central
Scandinavia. **Flowers** May–September. **In gardens**
Well-drained soil in partial shade. Rock gardens and wall
crevices. **Image size** Two-thirds life-size.

The ivy-leaved toadflax seems to be so at
home in the crevices of mossy old walls that
it is hard to believe that it only arrived
from Italy as an ornamental plant in the
17th century. Its tiny violet flowers have
two lips; the upper has two lobes, the lower
three plus a pale yellow patch that guides
pollinating insects to its nectar. As the
name suggests, the dark green leaves are
ivy-like and, like ivy, toadflax clings to walls
and rocks with its penetrating roots. It has
a cunning way of spreading its seed. When
toadflax is in bloom the flower stems lean
outwards towards the light. Once the
flowers have been fertilised, the stems turn
back on themselves so that they are
perfectly placed to lodge the ridged seeds
into crevices and cracks.

Brooklime [5]

Veronica beccabunga (Scrophulariaceae)

Aquatic perennial. Height Up to 60cm. **Habitat** Pond,
river and stream margins, marshes and fens. **Distribution**
All northern Europe. **Flowers** May–September. **In gardens**
Wet soil or shallow water. Wildlife ponds and bog
gardens. **Image size** Two-thirds life-size.

Hollow-centred stems that carry oxygen to
the roots help brooklime to survive its wet
habitat, where it is often found alongside
the celery-leaved buttercup (p.115). It is
related to speedwells, and has dense spikes
of blue flowers; the upper petal is slightly
larger than the three lower ones. The fleshy
oval leaves are toothed and are arranged in
opposite pairs up the stems. Despite their
bitter flavour, young brooklime leaves were
once added to salads and were boiled with

oranges and watercress to make a syrup for
preventing scurvy.

Heath dog violet [6]

Viola canina (Violaceae)

Perennial. Height Up to 30cm. **Habitat** Heaths, fens and
open woodland. **Distribution** Britain, northern Europe.
Flowers April–July. **In gardens** Moist, well-drained soil
in full sun or partial shade. Wild-flower and woodland
gardens. **Image size** Two-thirds life-size.

Like other dog violets, the flowers of this
species are unscented. It resembles the
common dog violet (see left), but the
dainty flowers, each of five petals, are pale
blue with a white or greenish-yellow spur.
The leaves are semi-evergreen and heart-
shaped but narrower than those of the
common dog violet; unlike the latter they
are not arranged in a rosette at the base of
the stem. The heath dog violet copes with
more sun than its relative.

Bittersweet, woody nightshade [2]
Solanum dulcamara (Solanaceae)

Scrambling perennial. Height Stems up to 3m long.
Habitat Damp woods, ditches, hedgerows, walls and
shingle beaches. **Distribution** All northern Europe except
far north. **Flowers** June–September. **Image size**
Two-thirds life-size.

Sometimes misnamed deadly nightshade
(p.217), all parts of this plant are poisonous
– especially the scarlet berries – but only
enough to cause vomiting. The species
name *dulcamara* means 'sweet-bitter'; the
first taste is intensely bitter, followed by a
sweet aftertaste.

Bittersweet is a familiar and flamboyant
plant of country lanes, where it scrambles
through hedgerows on wiry branched stems,
supporting a mass of arrow- or heart-shaped
leaves. Hanging in loose clusters, the
striking deep purple, starry flowers five
recurved petals with contrasting pale yellow
anthers fused together to form a distinctive
pyramid. In autumn, the clusters of berries
display a glorious spectrum of colours from
green through to yellow, orange and red. In
Germany, where bittersweet was hung
around the necks of cattle to ward off evil,
physicians employed the stems in
preparations to relieve rheumatism.

Zigzag clover [3]
Trifolium medium (Fabaceae or Leguminosae)

Perennial. Height Up to 50cm. **Habitat** Grassland and
scrub. **Distribution** All northern Europe except Faeroe Is
and Iceland. **Flowers** May–July. **Image size** Two-thirds
life-size.

The zig-zag clover is named for its stems,
which change angle slightly at each node.
This distinguishes it from related species
such as the red clover (p.160), which has
similar egg-shaped, pinkish-purple flower
heads. The foliage, typically for clovers, has
leaflets in threes (hence the botanical name
Trifolium). In this species the leaflets are
plain green and oval.

Jacob's ladder [1]
Polemonium caeruleum (Polemoniaceae)

Perennial. Height Up to 1m. **Habitat** Limestone grassland,
scree and rocky habitats. **Distribution** All northern
Europe except the far north and Ireland. **Flowers**
May–August. **In gardens** Well-drained, moist soil in full
sun or partial shade. Herbaceous borders, wild-flower
gardens and grassy sites. **Image size** Two-thirds life-size.

Jacob's ladder is a popular garden perennial
yet is one of Britain's rarest wild plants.

The charming flowers, held in upright
clusters, have five oval petals and five stamens,
which unzip to reveal vibrant orange pollen.
Up to 25 tiny, narrow leaflets, arranged in
pairs, supposedly look like the rungs of the
ladder that, according to the Old Testament,
Jacob dreamed the angels used for climbing
up to heaven. For a natural garden look,
plant Jacob's ladder or its white form var.
lacteum in grass; there, over time, its
creeping rootstock will create attractive drifts.

Meadow clary [4]

Salvia pratensis (Lamiaceae or Labiatae)

Perennial. Height 50–80cm. **Habitat** Chalky grassland, scrub and wasteland. **Distribution** Southern Britain, north to southern Sweden. **Flowers** June–July. **In gardens** Moist, well-drained soil in full sun or dappled shade. Mixed and herbaceous borders, and wild-flower and bee gardens. **Image size** Half life-size.

The name clary may be a corruption of 'clear-eye', from the time when its seeds were soaked in water for eye infections. Meadow clary is a faintly aromatic, hairy plant. It has upright, square stems that carry spikes of papery, violet-blue flowers in whorls. Typical of the sage family, the flowers consist of an upper and lower lip with two distinct, protruding stamens; the upper lip is curved. Plants have either male and female flowers, or only females, pollinated by bumblebees. Wrinkled, toothed leaves are arranged in pairs. Meadow clary is rare in the wild in Britain but has long been grown in gardens.

Columbine [5]

Aquilegia vulgaris (Ranunculaceae)

Perennial. Height 50–90cm. **Habitat** Woods, grassland and scrub, especially on chalk. **Distribution** All northern Europe except far north. **Flowers** May–July. **In gardens** Moist, well-drained soil in full sun or partial shade. Woodland gardens, herbaceous borders. **Image size** Half life-size.

Columbine's whimsical alias 'granny's bonnet' perfectly describes its nodding 3–5cm flowers. They have deep purple (or sometimes pink or white) rounded petals with hooked spurs at the base described by the herbalist John Gerard as 'little hollow hornes'. They dangle from upright, wiry stems and are pollinated by bumblebees. Each elegant leaf is made up of three leaflets with rounded lobes; they grow on stems that emerge at the base of the flower stalks. In its wild form, columbine has long been planted in cottage gardens, but it has also given rise to garden cultivars that range from the pure white 'Munstead White' to the decidedly brash 'Nora Barlow', with its mass of pale green and red pompons.

Wood crane's-bill [6]

Geranium sylvaticum (Geraniaceae)

Perennial. Height 30–70cm. **Habitat** Woods, meadows and hedge-banks. **Distribution** All northern Europe. **Flowers** June–July. **In gardens** Moist, well-drained soil in full sun or partial shade. **Image size** Half life-size.

The clump-forming wood crane's-bill has upright, branched flower stems with upward-facing, saucer-shaped 25mm flowers of five rounded petals; they vary from reddish-purple to blue. Whorls of leaves arranged up the stem consist of five to seven leaflets with deeply cut edges. Crane's-bills get their name from the long beak-like seed heads, which spring apart from the base when ripe. They make reliable, low-maintenance garden plants that thrive in most conditions.

Spring gentian [1]

Gentiana verna (Gentianaceae)

Perennial. Height Up to 7cm. **Habitat** Grassland, heaths and meadows. **Distribution** Northern England, western Ireland, central and eastern France and southern Germany. **Flowers** April–June. **In gardens** Moist, well-drained soil in full sun or partial shade. Rock and wild-flower gardens. **Image size** Two-thirds life-size.

The solitary sapphire-blue flowers of the very rare spring gentian grow on short tufted stems. They unfurl in late spring to reveal a white throat and stigma, and consist of a narrow tube up to 25mm long with five lobes at its tip. Unlike trumpet-shaped gentians, these lobes open so wide that they are nearly flat. A mat of bright green oval leaves grows in evergreen rosettes at the base of the stems. Sadly this beautiful gentian is rare and is protected by law in its few native sites, such as Upper Teesdale in England and the Burren in Ireland. However, it is available from nurseries.

Purple or blue gromwell [2]

Buglossoides purpurocaerulea (syn. *Lithospermum purpurocaeruleum*; Boraginaceae)

Perennial. Height Up to 60cm. **Habitat** Scrub, woodland and wasteland. **Distribution** Northern Europe as far north as southern Britain, Belgium and central Germany. **Flowers** April–June. **In gardens** Well-drained soil in full sun. Rock gardens, wild-flower gardens and borders. **Image size** Two-thirds life-size.

The rare purple gromwell is recognisable by its flowers, which open reddish-purple and mature a startling gentian-blue. They grow in spiralled cymes – although less obviously than other members of the borage family, such as viper's bugloss (p.176) – which nestle among hairy, lanceolate leaves. Each flower consists of a petal tube with five lobes and is followed by a tiny nutlet. Purple gromwell is available from a number of nurseries. Picking the flowers is a waste of time; once in water the petals turn dull blue and the leaves shrivel.

Salsify [3]

Tragopogon porrifolius (Asteraceae or Compositae)

Biennial. Height Up to 1m. **Habitat** Grassy and wasteland. **Distribution** Introduction. Most of northern Europe to southern Sweden. **Flowers** May–August. **Image size** Two-thirds life-size.

Introduced to northern Europe from the Mediterranean for its edible fleshy root and fresh young shoots, salsify colonised tracts of wasteland. Its solitary lilac flower heads are similar to those of the related yellow goat's-beard (p.95) but are larger, with strap-shaped ray florets and long, slim, pointed bracts; they close up at midday. The large dandelion-clock seed heads consist of a mass of individual seeds each with silky hairs that carry it on the wind. The long, narrow leaves are hairless, with smooth edges.

Spring squill [4]

Scilla verna (Liliaceae)

Bulbous perennial. Height Up to 15cm. **Habitat** Coastal grassy and rocky habitats. **Distribution** Britain, Ireland, Faeroe Is, France and Norway. **Flowers** April–June. **Image size** Two-thirds life-size.

The spring squill produces squat racemes of starry flowers, in violet-blue and occasionally

white, on erect stems. The foliage is spindly and grassy, and grows from the bulbs. The spring-flowering squill – *verna* means 'of spring' – thrives in short grass on cliff tops on the west coast of Britain and elsewhere.

Bugle [5]

Ajuga reptans (Lamiaceae or Labiatae)

Creeping perennial. Height Up to 30cm. **Habitat** Woodland, damp grassland and wasteland. **Distribution** All northern Europe. **Flowers** April–June. **In gardens** Moist, well-drained soil in full sun or partial shade. Borders (especially where shady) and woodland gardens. **Image size** Two-thirds life-size.

A common native in many parts of Europe, bugle is better known as an ornamental plant than a wild flower. Its upright stems grow from vigorous creeping rhizomes, with leafy spikes of dark velvety-blue, 15mm tubular flowers. Each has two lips – the upper almost invisible, the lower with three lobes. The leaves, in opposite pairs, are oval and dark green; they may be flushed with bronze. Bugle's basal leaves form dense mats – good for ground cover, but it can be invasive.

Pasque flower [6]

Pulsatilla vulgaris (syn. *Anemone pulsatilla*; Ranunculaceae)

Perennial. Height Up to 30cm. **Habitat** Dry grassland, often on chalk or limestone. **Distribution** Britain, Belgium, Holland, France, Germany, Denmark and southern Sweden. **Flowers** April–June. **In gardens** Well-drained soil in full sun. Rock gardens and dry borders. **Image size** Two-thirds life-size.

Pasque flower is a shy yet exotic beauty, now rarely seen in the wild. The name refers to Easter, when the large cup-shaped flowers – up to 85mm across – unfurl on short stems from their hairy buds. They have five or six violet-purple petaloid sepals and an explosion of vibrant yellow stamens at their centre. The leaves are bipinnate and feathery, and the plant is covered in a fuzz of hairs.

Hedge woundwort [2]

Stachys sylvatica (Lamiaceae or Labiatae)

Creeping perennial. Height 60–100cm. **Habitat** Woods and hedgerows. **Distribution** All northern Europe except Faeroe Is and Iceland. **Flowers** June–September. **In gardens** Well-drained soil in full sun or partial shade. Wild-flower gardens. **Image size** Two-thirds life-size.

The genus name *Stachys* comes from the Greek for 'ear of corn' and refers to the corn-like appearance of the flower heads. The common name woundwort has greater significance because the plant and its relatives have legendary healing properties that go back to the 16th century; tests have proved that it contains a mild antiseptic. Hedge woundwort is a robust plant that spreads on tough, hairy, creeping stems. The flower spikes consist of dull reddish-purple flowers, each 12–18mm long, with a hairy upper and lower lip, and are arranged in whorls. White markings on the lower lip guide pollinating bees to the nectar. The heart-shaped leaves are also covered in hairs and have toothed edges. The plant gives off a strong unpleasant smell.

Viper's bugloss [3]

Echium vulgare (Boraginaceae)

Biennial. Height 50–100cm. **Habitat** Grassland, dunes and wasteland, especially on chalky soil. **Distribution** All northern Europe except Faeroe Is and Iceland. **Flowers** June–September. **In gardens** Well-drained soil in full sun. Wild-flower gardens and annual borders. **Image size** Two-thirds life-size.

Viper's bugloss has several associations with snakes. The spent flower heads and their nutlets were thought to resemble a snake's head and to 17th-century herbalists the speckled stems looked like a snake's skin, so viper's bugloss was believed to have healing properties for snakebite.

It is a striking plant whose floral display begins with pinkish buds held on short, curved, hairy sprays that look like exotic caterpillars. Then the violet-blue, funnel-

Sheep's-bit, sheep's-bit scabious [1]

Jasione montana (Campanulaceae)

Annual or biennial. Height 20–50cm. **Habitat** Dry grassland, heaths, cliffs and shingle. **Distribution** All northern Europe except Faeroe Is and Iceland. **Flowers** May–August. **In gardens** Well-drained, sandy soil in sun. **Image size** Two-thirds life-size.

Sheep's-bit looks uncannily like devil's-bit scabious (p.182), both having round heads of numerous flowers with prominent stamens. In fact, sheep's-bit belongs to the bellflower family and prefers very different growing conditions to devil's-bit scabious. Its individual flowers each have five linear, pale blue petals; the entire globe is supported by a ruff of bracts. The leaves of the two plants are quite different. In sheep's-bit they are hairy, lanceolate, slightly wavy and arranged in a spiral along the the stems.

shaped 20mm flowers emerge, in spikes that make attractive pyramids of colour. Finally, as the seed heads develop, the sprays start to lengthen. The elliptical leaves are covered in a mass of rough white hairs and are arranged in a rosette at the base of the stem; bugloss comes from the Greek for 'ox-tongue' and describes their shape.

Purple toadflax ④

Linaria purpurea (Scrophulariaceae)

Perennial. Height Up to 1m. **Habitat** Waste and cultivated land and old walls. **Distribution** Naturalised, mainly in Britain and Ireland. **Flowers** June–August. **In gardens** Well-drained soil in full sun. Dry borders and wild-flower gardens. **Image size** Two-thirds life-size.

The slender, diaphanous spires of purple toadflax may seem at home on old walls and railway embankments in England, but it is actually native to Italy, where it thrives in dry heat on stony and sandy soils. The purple-violet snapdragon-like flowers each have two lips and a long, curved spur where nectar is stored. The narrow leaves are in whorls at the base of the stem but are arranged alternately farther up. Purple toadflax is a favourite, particularly for dry or gravel gardens. It self-seeds freely, so weed out seedlings to keep it under control.

Marsh violet ⑤

Viola palustris (Violaceae)

Creeping perennial. Height Up to 10cm. **Habitat** Marshes, bogs, wet heathland and damp woodland, especially at altitudes of 1,200–2,600m. **Distribution** Throughout northern Europe. **Flowers** April–July. **Image size** Life-size.

The marsh violet is stemless; the kidney-shaped leaves and delicate lilac flowers arise directly from creeping underground rhizomes. This lack of stem is the easiest way to distinguish it from other wild violet species. The flowers have five petals decorated with a tracery of dark purple veins and a

short spur, but are not scented. The leaves are larger than those of most violet species.

Heath speedwell ⑥

Veronica officinalis (Scrophulariaceae)

Perennial. Height Up to 40cm. **Habitat** Grassland, heaths and open woodland. **Distribution** All northern Europe. **Flowers** May–August. **Image size** Life-size.

Heath speedwell produces spikes of pretty, lilac-blue small flowers with distinctive dark veins, a white eye and two prominent stamens. It is a compact plant, the oval leaves growing on short stalks and arranged in pairs on creeping stems. It is usually hairy, but hairless forms appear occasionally. The flowers are pollinated by flies and bees, and are followed by hairy heart-shaped fruits.

Honesty [1]

Lunaria annua (Brassicaceae or Cruciferae)

Biennial. Height Up to 1m. **Habitat** Waste and cultivated land and roadsides. **Distribution** Introduction. Naturalised in most except far northern Europe. **Flowers** April–June. **In gardens** Moist, well-drained soil in full sun or partial shade. Wild-flower and woodland gardens, and mixed borders. **Image size** One-and-a-half life-size.

Originally from south-eastern Europe, honesty is all but a northern European native, having happily naturalised along railway embankments and roadsides. It is an upright, hairy plant with heart-shaped toothed leaves. The romantic cottage-garden flowers are held in a loose inflorescence; each has four petals in shades of reddish-purple or occasionally white. They are followed by almost circular, flattened fruits, 30–55mm across, which split when ripe to expose the silvery middle wall, or septum. These (see photograph right) are greatly prized by flower arrangers and have earned honesty the common names silver dollar, money plant and moonplant. Planted in a wild-flower garden, honesty is a magnet for butterflies and bees. *Lunaria annua* 'Variegata' has attractive heart-shaped foliage with a creamy-white margin.

Bugloss [2]

Anchusa arvensis (syn. *Lycopsis arvensis*; Boraginaceae)

Annual. Height Up to 50cm. **Habitat** Arable, waste and bare land. **Distribution** All northern Europe except Iceland and far north. **Flowers** May–September. **Image size** Life-size.

The widespread and increasing use of fertilisers and herbicides has caused the rapid demise of bugloss over the last 50 years. It is, perhaps, a plant that might not even merit a second glance from a casual observer. The upright stems are stiff and covered in bristles. The oval leaves are wavy-edged, with the lower ones arranged in a rosette and the upper clasping the stem; they are also bristly, and in shape and texture are supposedly similar to an ox's tongue. Tiny forget-me-not-like flowers appear at the tip of the stems; they are bright blue with a white centre, and are held in curved sprays on stems that lengthen as the fruits ripen. In some species, the roots yield a red extract that was used to stain wood and cosmetically as rouge.

Southern marsh orchid [3]

Dactylorhiza praetermissa (Orchidaceae)

Tuberous perennial. Height Up to 60cm. **Habitat** Fens, marshy areas and wet meadows. **Distribution** Most of northern Europe as far north as southern Scandinavia. **Flowers** June–August. **Image size** Life-size.

Similar in shape to the closely related common spotted orchid (p.131), this species usually has unspotted leaves, which are flat or slightly hooded at the top. The flowers are reddish-purple. The lip is usually quite broad and at most notched rather than lobed, and is marked with dots and lines in the centre. The spur is stout and conical. Southern marsh orchid will hybridise with at least four other *Dactylorhiza* species, forming a complicated group. The occasional plant may have ring-shaped leaf spots. These were originally thought to be the result of crossing with the common spotted orchid, but chromosome studies show that the spotted variety is distinct from the true hybrid.

Early purple orchid [4]

Orchis mascula (Orchidaceae)

Tuberous perennial. Height Up to 40cm. **Habitat** Woods, grassland, road verges and scrub. **Distribution** All northern Europe except Iceland. **Flowers** April–June. **Image size** One-quarter life-size.

As beautiful as the flowers of the early purple orchid are, their odour – which begins sweetly enough – is said to be unpleasantly reminiscent of tomcats as the flowers fade. This species is rather like a larger version of the green-winged orchid (see below), with its thick spikes of purple flowers that occasionally appear pinkish or white. The upper sepal and petals are fused together to form a hood, while the lower lip has three lobes and central spots to attract the pollinating insects that use it as a landing platform. Unlike the green-winged orchid, the surface of the lance-shaped leaves is covered in attractive dark purple spots.

Early purple orchid has, for thousands of years, been associated with sex. The ancient Greeks used it to predict the sex of their children, and in Ireland it was the basic ingredient of a powerful love potion.

Green-winged orchid [5]

Orchis morio (Orchidaceae)

Tuberous perennial. Height Up to 20cm. **Habitat** Grassy habitats. **Distribution** All northern Europe except Faeroe Is, Iceland, Finland and far north. **Flowers** May–June. **Image size** Half life-size.

This orchid needs undisturbed short grassland to thrive, so it is sadly declining along with its habitat as a result of intensive farming methods. It is similar to the early purple orchid (see above) but is smaller and has no purple blotches on its leaves. The flowers consist of sepals and petals that together form a hood; the lower lip has three shallow lobes. They vary in colour from purple through reddish to pink, or are occasionally whitish. The green wings of the name are the lateral sepals with their distinctive fine green veins. The foliage is lanceolate, with the lower leaves arranged in a rosette and the upper sheathing the lower part of the stem. In the 16th century the hood-like appearance of the flowers was believed to have a certain sexual symbolism.

Wild clary [1]

Salvia verbenaca (Lamiaceae or Labiatae)

Perennial. Height Up to 80cm. **Habitat** Grassland, roadsides and dunes. **Distribution** Britain and France. **Flowers** May–August. **Image size** Two-thirds life-size.

Wild clary produces whorls of flowers in shades of blue, violet or lilac that, en masse, form a loose spike. Each flower has two lips – the upper hooded and the lower three-lobed with white markings to entice insects. Some of the flowers do not open and are self-fertilising. They grow on square, branched stems that are tinged with red. The large, oblong, wrinkled leaves have toothed edges; they grow from the base of the stem on long stalks. Like the rare meadow clary (p.173), the seeds were used to make an eyewash.

Perennial flax [2]

Linum perenne (Linaceae)

Perennial. Height 30–60cm. **Habitat** Chalky grassland. **Distribution** Parts of Britain, eastern France and southern Germany. **Flowers** June–July. **In gardens** Well-drained soil in full sun. Wild-flower and herbaceous borders. **Image size** Two-thirds life-size.

Perennial flax is now so scarce in the wild that sighting this pretty plant is unforgettable. The glossy, funnel-shaped, clear blue 20–25mm flowers unfurl from small spiral buds on upright, wiry stems. They have five petals and last only a day, their appearance usually over by late afternoon. Masses of long, narrow blue-green leaves grow up the length of the stems.

Blue fleabane [3]

Erigeron acer (Asteraceae or Compositae)

Annual or biennial. Height Up to 50cm. **Habitat** Sandy soil, dry grassland and disturbed ground, often on chalk. **Distribution** All northern Europe except Faeroe Is and Iceland. **Flowers** July–September. **Image size** Two-thirds life-size.

Blue fleabane is a master at colonising disturbed soil. After flowering, it releases hundreds of seeds, which are carried on the wind. They can produce two generations of seedlings – those that flower the same year (annuals) and, from those that germinate later, biennials that flower early the following year. The flowers themselves have pale purple-mauve ray florets – not blue – with a yellow cone of disc florets in the centre; they grow, one per branch, on lanky, purplish stems. The oval hairy leaves may be toothed or untoothed; they grow in a rosette at the base and alternately up the length of the stems. Fleabanes got their name because they were burned to repel fleas.

Russian comfrey [4]

Symphytum x *uplandicum* (Boraginaceae)

Perennial. Height Up to 2m. **Habitat** Woodland, roadsides and other wasteland. **Distribution** Introduction to all northern Europe. **Flowers** June–August. **Image size** Two-thirds life-size.

Russian comfrey is the result of a cross between the white, purplish or pinkish common comfrey (p.76) and its blue-flowered cousin rough comfrey (*Symphytum asperum*). It is now more common in many areas than either. It has roughly hairy oval leaves and stems. Comfrey leaves are said to be delicious dipped in batter and fried. Steeped in water they make a high-potash feed for almost any plant.

Borage [5]

Borago officinalis (Boraginaceae)

Annual. Height 30–60cm. **Habitat** Waste and arable land, sandy heaths. **Distribution** Naturalised in most of northern Europe except Ireland, Iceland and Scandinavia. **Flowers** May–September. **In gardens** Well-drained soil in full sun or partial shade. Dry gardens. **Image size** Half life-size.

Borage originated in the Mediterranean region and plants that colonise wild habitats farther north escaped from gardens. It is a bristly plant with upright stems and toothed, wavy-edged, lanceolate leaves. The bright blue star-like flowers have blackish stamens in a distinctive spike. Borage has long been cultivated for its cucumber-like taste.

Hound's-tongue [6]

Cynoglossum officinale (Boraginaceae)

Biennial. Height Up to 60cm. **Habitat** Dry grassland, wood edges and dunes. **Distribution** All northern Europe except Iceland and far north. **Flowers** May–August. **Image size** One-quarter life-size.

Hound's-tongue is so named because the hairy, grey, lanceolate leaves were thought to look rather like a dog's tongue. What's more, they were used to soothe bites from rabid dogs. In fact, the plant smells of mice, which is strong enough to repel anyone, let alone a dog on the rampage. The flowers are a deep but dull purple, and funnel-shaped, with five rounded petals. The fruits that follow contain four nutlets covered in hooked spines that attach to the fur of passing animals. As they ripen, the stems lengthen and become more curved.

Spanish bluebell [7]

Hyacinthoides hispanica (syns. *Endymion hispanicus*, *Scilla hispanica*; Liliaceae)

Bulbous perennial. Height Up to 40cm. **Habitat** Woods, copses and banks. **Distribution** Introduction to Britain and France. **Flowers** May–June. **In gardens** Well-drained soil in dappled shade. Woodland gardens and shady borders. **Image size** One-third life-size

The Spanish bluebell, from Portugal and Spain, has made a highly successful assault on the north-western European native bluebell (p.170) – not least because it tolerates a wide range of soil types and spreads vigorously and quickly. It is distinguished from the native bluebell by having racemes of flowers that are more upright, with the flowers carried on both sides. The flowers are a wider bell shape, with slightly flared edges. Unlike the native bluebell, they are not scented. The foliage is strap-like, glossy and dark green.

Round-headed rampion [1]

Phyteuma orbiculare (Campanulaceae)

Perennial. Height Up to 50cm. **Habitat** Chalky grassland, embankments and rocky habitats. **Distribution** Britain, France and Germany. **Flowers** June–August. **Image size** Two-thirds life-size.

British poet and naturalist Geoffrey Grigson aptly described the round-headed rampion as a 'violet sea-anemone [or] air-anemone'. The solitary, dense, globular flower heads are held aloft on erect grooved stems and consist of corollas that curve inwards, just like Grigson's exotic anemone. The heart-shaped leaves are deeply veined and may have rounded teeth.

Skullcap [2]

Scutellaria galericulata (Lamiaceae or Labiatae)

Creeping perennial. Height Stems up to 50cm long. **Habitat** Wet meadows, fenland, pond edges and river-banks. **Distribution** All except far northern Europe. **Flowers** June–September. **In gardens** Well-drained soil in full sun or partial shade. Wild-flower gardens or containers. **Image size** Two-thirds life-size.

Skullcap is named after Roman legionnaires' leather helmets, which the calyx supposedly resembles. The plant contains the volatile oil scutellarin, and the dried flowers and leaves were used for relieving depression, anxiety and hysteria. The tiny purple-blue flowers have two lips; they appear up the stem at the nodes of oval (or sometimes heart-shaped) leaves.

Wild pansy, heartsease [3]

Viola tricolor (Violaceae)

Annual or short-lived perennial. Height 10–30cm. **Habitat** Grassland, wasteland and cultivated land. **Distribution** All northern Europe. **Flowers** April–October. **In gardens** Moist, well-drained soil in full sun, partial shade. Rock gardens and lawns. **Image size** Two-thirds life-size.

A long association with love gave *Viola tricolor* the names heartsease and love-in-idleness. 'Pansy' comes from the medieval French *pensée* (thought). The charming five-petalled, spurred flowers look like miniature faces – with numerous combinations of purple, yellow and white face-paint. The oval leaves have blunt teeth.

Self-heal [4]

Prunella vulgaris (Lamiaceae or Labiatae)

Perennial. Height Up to 30cm. **Habitat** Grassland; open woodland. **Distribution** All except far northern Europe. **Flowers** June–November. **In gardens** Well-drained soil in sun or partial shade. Ground cover, lawns and containers. **Image size** Two-thirds life-size.

Self-heal's dark violet flowers have an upper and lower lip, the hooked upper lip supposedly like a sickle; so, according to the 16th-century doctrine of signatures, it was believed to mend wounds from sickles and billhooks. The flower heads, arranged in dense whorls, nestle on pairs of bright green, stemless leaves. They are pollinated by long-tongued bees delving for nectar. After the petals have fallen the calyxes that remain look like an ear of corn.

Wild thyme [5]

Thymus polytrichus (syn. *T. praecox*; Lamiaceae or Labiatae)

Creeping perennial. Height Stems up to 10cm long. **Habitat** Sandy and rocky grassland and heaths. **Distribution** All northern Europe except Denmark, Finland and Sweden. **Flowers** May–September. **In gardens** Well-drained soil in full sun. Rock gardens and paving; wild-flower gardens. **Image size** Two-thirds life-size.

Wild thyme's warm, spicy scent hangs on the summer air, attracting nectar-seeking bees. Mats of short, hairy stems are tipped with rounded spikes of pinkish-purple flowers. The small oval leaves grow in opposite pairs; tiny glands contain the fragrant, antiseptic oil thymol. Like those of the bushier garden thyme (*T. vulgaris*), the leaves are used in cooking.

Devil's-bit, devil's-bit scabious [6]

Succisa pratensis (Dipsacaceae)

Perennial. Height 50–100cm. **Habitat** Grassy habitats. **Distribution** All northern Europe. **Flowers** July–October. **In gardens** Moist soil in full sun or partial shade. Damp wild-flower gardens and meadows. **Image size** Two-thirds life-size.

According to a 15th-century story, the devil was furious at this plant's powerful medicinal properties, and bit off the roots in a fit of pique – hence the stubby rootstock. The 15–25mm flower heads – scabious-like, with prominent stamens, but more rounded – are held on upright, wiry stems. They vary from lilac to dark purple. Devil's-bit used to be classified as a true scabious (p.185 & 193), but is now placed in a separate genus in the same family. (Sheep's-bit, p.176, looks similar but is unrelated.) Close inspection of the flowers shows why: they have four lobes rather than five like true scabious. The elliptical leaves are arranged in pairs, the upper ones often toothed.

Lucerne, alfalfa [7]

Medicago sativa (Fabaceae or Leguminosae)

Perennial. Height 40–90cm. **Habitat** Roadsides, grassland, wasteland and cultivated land. **Distribution** Naturalised in all northern Europe except Faeroe Is and Iceland. **Flowers** June–July. **Image size** Two-thirds life-size.

Alfalfa is probably best known for its whispy, nutty-flavoured seedlings, which are a tasty and fashionable salad ingredient. Farmers grow it for fodder and to enrich the soil. (Like other members of the clover family, its root nodules contain specialised bacteria that 'fix' nitrogen from the air as nitrate, which plants use for food.) The leaves, which consist of three oval leaflets with toothed edges, look very similar to clover foliage. The dainty flower heads consist of dense clusters of deep purple to occasionally white, pea-like flowers. The variety of farm-cultivated forms means that escapees vary widely in leaf size and flower colour.

Creeping bellflower [1]

Campanula rapunculoides (Campanulaceae)

Spreading perennial. Height 30–80cm. **Habitat** Hedgerows, meadows, woodland and rocky habitats. **Distribution** All except far northern Europe; naturalised in Britain and Ireland. **Flowers** July–September. **Image size** Two-thirds life-size.

Whether native or a garden escapee, creeping bellflower grows in a wide variety of habitats – usually in force, forming impressively large patches. The pendulous, violet-blue bell-shaped flowers have five lobes. Arranged along the length of reddish-brown stems, each flower is partnered by a single oval leaf. The genus name *Campanula* comes from the Latin *campana* (bell), while *rapunculoides* suggests that the turnip-shaped roots resemble those of the rampion bellflower, *C. rapunculus*.

Small scabious [2]

Scabiosa columbaria (Dipsacaceae)

Perennial. Height 10–70cm. **Habitat** Grassy and rocky habitats. **Distribution** Most of northern Europe, as far north as southern Sweden, except Ireland. **Flowers** July–August. **In gardens** Well-drained soil in full sun. Rock gardens, wild-flower gardens and herbaceous borders. **Image size** Two-thirds life-size.

The petite 2–4cm flower heads of small scabious are like bright lilac-blue buttons that appear to hover on wiry, branched stems, well above the grey-green, downy foliage. Each consists of a mass of minute five-lobed flowers, those on the outside being larger than the ones towards the centre. The flower heads are more domed than those of field scabious (p.193). The oval basal leaves may or may not be lobed, but farther up the stem they are much more divided and feathery. The Latin name *Scabiosa* is less glamorous – it means 'itch', as with the disease scabies. Herbalists used scabious, dried and as a juice, to cure skin complaints. 'Pink Mist' and 'Butterfly Blue' are two cultivars worthy of garden planting.

Common, black or lesser knapweed [3]

Centaurea nigra (Asteraceae or Compositae)

Perennial. Height Up to 1m. **Habitat** Meadows, roadsides and railway embankments. **Distribution** All northern Europe to central Scandinavia. **Flowers** June–September. **Image size** Two-thirds life-size.

Called 'common' for good reason, this plant grows happily in challenging environments, spread on the wind by its many seeds, which have parachutes of tiny hairs. The flower heads consist of numerous purple florets, which are held by a corset of feathery bracts. The country name hardheads refers to the flower heads; knap is another word for 'knob', while the species name *nigra* refers to the blackish bracts. The narrow leaves grow alternately up the stems. Knapweed's medicinal uses include relieving sore throats and healing wounds and scabs. A woman would pluck the opened florets from a flower head, then place it inside her blouse. If any florets opened, a lover would be coming her way.

Melancholy thistle [4]

Cirsium helenioides (syn. *C. heterophyllum*; Asteraceae or Compositae)

Spreading perennial. Height Up to 1.2m. **Habitat** Damp grassland, woodland and scrub. **Distribution** Most of northern Europe except Belgium, Holland and Faeroe Is; naturalised in Iceland. **Flowers** July–September. **Image size** Two-thirds life-size.

In bud, the flower heads of this thistle droop in an apparently sad fashion, leading to both its common name and its medicinal use. This, according to the 17th-century herbalist Nicholas Culpeper, was to 'expel superfluous Melancholy out of the body'. The purple 3–5cm flower heads are either arranged in clusters or appear alone on upright stems – which, unlike other thistles, have no prickles. The leaves have soft prickly margins and undersides covered in white hairs, giving it another British country name, fish-belly.

Purple loosestrife [5]

Lythrum salicaria (Lythraceae)

Perennial. Height Up to 1.5m. **Habitat** Freshwater margins, marshes and fens. **Distribution** All northern Europe except Faeroe Is and Iceland. **Flowers** June–August. **In gardens** Moist soil in full sun. Bog gardens, pond margins and damp borders. **Image size** Two-thirds life-size.

Purple loosestrife's stout, upright stems carry dense whorls of reddish-purple flowers. The overall effect is of a mass of vibrant candle-like spikes against a background of dark green foliage. Each starry flower has six petals and 12 stamens. However, they can differ from one plant to another, there being up to three types of flower – a phenomenon known as trimorphism. Each has different-sized stamens, styles and pollen grains – a device to prevent self-pollination. They are pollinated by long-tongued bees and other insects, and sometimes caterpillars. There are several garden cultivars, including 'Feuerkerze' (syn. 'Firecandle'), with rose-red flowers.

Bloody crane's-bill [6]

Geranium sanguineum (Geraniaceae)

Perennial. Height 10–40cm. **Habitat** Grassland, rocky habitats, open woodland and dunes. **Distribution** All northern Europe except Iceland and Faeroe Is. **Flowers** July–August. **In gardens** Well-drained soil in sun or partial shade. Herbaceous or mixed borders; wild-flower gardens. **Image size** Two-thirds life-size.

Like many wild true geraniums, the bloody crane's-bill has made the transition from wild flower to ornamental garden plant with ease, partly because they happily perform with minimum maintenance. This species produces bright reddish-purple, papery saucer-like flowers, up to 4cm across, with dark veins and a white eye at their centre; the petals are notched. The solitary flowers are followed by elongated, beak-like seed heads, which give the plant its common name. The rounded foliage is deeply cut, giving an overall feathery effect.

Chives [1]

Allium schoenoprasum (Liliaceae)

Perennial. Height Up to 40cm. **Habitat** Damp grassy and rocky habitats, usually on alkaline soil. **Distribution** All northern Europe except Faeroe Is and Iceland. **Flowers** June–August. **In gardens** Any well-drained moist soil in sun or partial shade. Herb gardens and front of borders; also as companion plant against apple scab and rose black spot. **Image size** Half life-size.

Chives are more often seen in the kitchen, chopped up in salads, than in the wild. you come across a plant with long hollow leaves and a stem topped by a dense round head of pinkish-purple flowers with six segments, crushing a leaf will quickly identify it. The English name is derived from the old French *cive*, itself derived from the Latin *cepa* (onion). In the garden, however, chives can be harvested up to four times a year.

Common cornsalad, lamb's lettuce [2]

Valerianella locusta (Valerianaceae)

Annual. Height Up to 15cm. **Habitat** Cultivated and wasteland, sparse grassland, cliffs and dunes. **Distribution** All northern Europe, especially near coasts in far north. **Flowers** April–June. **Image size** Half life-size.

The spoon-shaped leaves of cornsalad were enthusiastically eaten as a salad vegetable in France long before they became popular in Britain. They form a rosette at the base of the prickly, branched stems, with more oval to lanceolate leaves in opposite pairs farther up. Long, wiry flower stems grow from the leaf axils and produce tight flattish umbels of tiny lilac flowers with five petals. A distinctive ruff of bracts envelops each umbel. The fruits are small and rounded, only 2mm across. However, several closely related species exist whose only distinguishing features are their fruits – narrow-fruited, broad-fruited and keel-fruited cornsalads (*V. dentata*, *V. rimosa* and *V. carinata* respectively). Cornsalad is easy to grow for salads on a 'cut-and-come-again' basis.

Tufted vetch [3]

Vicia cracca (Fabaceae or Leguminosae)

Scrambling or climbing perennial. Height Stems up to 2m long. **Habitat** Grassy habitats, hedgerows, scrub and embankments. **Distribution** All northern Europe. **Flowers** June–August. **Image size** Two-thirds life-size.

The tufted vetch uses long, branched tendrils that grow at the tips of its leaves to support itself as it clambers through hedgerows and along embankments. Up to 40 pea-like flowers are arranged in dense one-sided racemes and appear in shades of bluish-violet. The flowers are followed by smooth, brown seedpods, which each carry up to six seeds. The foliage is finely divided, each leaf consisting of 12 to 30 linear leaflets in pairs; they form a ladder-like effect similar to Jacob's ladder (p.172).

Sand leek [4]

Allium scorodoprasum (Liliaceae)

Perennial. Height Up to 80cm. **Habitat** Grassland and scrub. **Distribution** All northern Europe. **Flowers** June–August. **Image size** Two-thirds life-size.

The sand leek isn't blessed with the good looks possessed by many of the ornamental species of onions grown, because its flower heads produce only a smattering of flowers. The bell-shaped flowers grow on fine stalks of varying lengths, and vary in colour from pale lilac through to purple; they are followed by clusters of shiny purple bulbils. Blue-grey linear leaves envelop the upright stems. As might be expected from a close relative of garlic and chives, the whole plant exudes a strong aroma and flavour. It has been cultivated for culinary purposes.

Marsh thistle [5]

Cirsium palustre (Asteraceae or Compositae)

Biennial. Height 1–2m. **Habitat** Marshes, damp grassland and woods. **Distribution** All northern Europe except Iceland. **Flowers** July–September. **Image size** Two-thirds life-size.

Wherever there are damp conditions the marsh thistle is bound to grow. It is a common plant, but only one of about 100 related thistle species. The curvaceous 1–2cm flower heads produce plumes of tubular purple florets that appear to explode from the corset of overlapping bracts underneath. Once the flowers have been pollinated by bees and butterflies, the bracts dry out and open up to set the seeds free. These are dispersed on the wind by parachutes of silky hairs. In contrast to its elegant blooms, the marsh thistle has tall stems with spiny wings along their length. The pointy leaves are also equipped with sharp spines. The welted thistle (*Carduus crispus*, syn. *C. acanthoides*) is similar, but the spiny wings stop just below the flower heads – which are a more reddish-purple.

Giant bellflower [6]

Campanula latifolia (Campanulaceae)

Perennial. Height Up to 1.2m. **Habitat** Damp woodland. **Distribution** Britain, France, Germany, Denmark, Norway, Sweden and Finland. **Flowers** July–August. **In gardens** Moist, well-drained soil in partial shade. Wild-flower gardens, herbaceous borders. **Image size** Two-thirds life-size.

This, the largest and most robust member of the *Campanula* genus, produces crowded racemes of impressive bell-shaped flowers up to 55mm long on hairy and slightly angled stems. They range from pale purple through to almost white. The plant closely resembles the nettle-leaved bellflower (*C. trachelium*) except that all its parts are larger. Each bell has five lobes and long leafy sepals; although the flowers turn upwards, the seed capsules that follow hang down. The oval leaves are toothed (more evenly so than in *C. trachelium*); at the base of the stem they have winged stalks, but farther up they are more or less stalkless. Giant bellflower grows as vigorously as it is robust.

Foxglove [1]

Digitalis purpurea (Scrophulariaceae)

Biennial or short-lived perennial. Height Up to 1–2m. **Habitat** Woodland clearings, heaths and wasteland on acid soils. **Distribution** All northern Europe to central Sweden. **Flowers** June–September. **In gardens** Most soils in partial shade. Woodland gardens and shady borders. **Image size** Half life-size.

The foxglove has become a stalwart of cottage gardens with its statuesque spikes of supposedly glove-shaped flowers. The purple corolla has two lips; the bottom lip is slightly hairy both on the edge and inside. Up to 80 pendulous blooms, each up to 55mm long, may appear on one stem. The leaves are oval and grow in opposite pairs along the stem; at the base they form a rosette and are slightly toothed.

The foliage can be deadly poisonous, probably accounting for why in Ireland it is known as fairy thimbles, in Wales as elves' fingers or elves' gloves. Magic aside, in the late 18th century it was found to be effective in controlled doses for treating heart conditions. It contains digoxin and digitoxin, still widely used for this purpose.

Sea pea [2]

Lathyrus japonicus (Fabaceae or Leguminosae)

Perennial. Height Up to 90cm. **Habitat** Coastal shingle, sand dunes and rocks; sometimes inland lake margins. **Distribution** North-western Europe, from Britain and Ireland to Scandinavia. **Flowers** June–August. **Image size** Twice life-size.

Despite its botanical name, this low-growing, sprawling member of the pea family is a European native plant. It is a ubiquitous coloniser of shingle beaches, the stems creeping along the shoreline, conspicuous for their closely-set pairs of oval, blue-green and rather fleshy leaflets. Short, upright flower stems emerge from the leaf axils, with clusters of up to ten prominent pea flowers at their tip. These are 18–22mm long and start life a vibrant purple, eventually fading to blue.

Common butterwort [3]

Pinguicula vulgaris (Lentibulariaceae)

Perennial. Height Up to 15cm. **Habitat** Bogs, moors, damp heaths and wet rocks. **Distribution** Most of northern Europe except much of southern England and southern Ireland. **Flowers** May–July. **Image size** Half life-size.

The solitary blooms of common butterwort grow on slim, upright stems and are seductive enough to tempt hapless insects into the plant's carnivorous clutches. They are violet with a white patch in the throat, and have two lips, both deeply lobed, and a long spur behind. The yellowish-green oblong leaves are arranged in a star-like rosette at the base of the flower stems. Each leaf has curled edges and a sticky surface. When an insect lands on a leaf it sticks fast, and, struggling to escape, causes the leaf to fold up around it. The leaf exudes enzymes to digest the soft parts of the insect's body, from which the plant absorbs valuable nutrients. Butterwort was believed to protect cattle from elves' arrows and newborn babies from thieving fairies. The leaves were used to curdle or thicken milk.

Bush vetch [4]

Vicia sepium (Fabaceae or Leguminosae)

Climbing perennial. Height Up to 60cm. **Habitat** Grassy habitats, hedgerows, scrub and woodland. **Distribution** All northern Europe except Faeroe Is. **Flowers** May–November. **Image size** Half life-size.

Like tufted vetch (p.186), bush vetch clambers its way through supporting plants by clinging on with winding tendrils at the tips of the leaves. The purplish-blue flowers have five lobes, which only bumblebees can prise apart for their nectar. They grow on short stems in clusters of two to six flowers and are followed by hairless, black seedpods. The leaves consist of between three and nine pairs of oval leaflets, which create an attractive ladder-like effect – again like tufted vetch, except that the leaflets have a broader appearance.

Venus's looking glass [5]

Legousia hybrida (Campanulaceae)

Annual. Height Up to 30cm. **Habitat** Arable land and bare, stony ground on chalky soil. **Distribution** Most of northern Europe except Faeroe Is, Iceland and Scandinavia. **Flowers** May–August **Image size** Two-thirds life-size.

An elusive cornfield annual that has suffered from 'improvements' to the agricultural land that is its natural habitat, Venus's looking glass is related to the harebell (p.193) and clustered bellflower (p.190). It is, however, quite different from its cousins, with tiny purple star-like flowers that perch on top of a distinctive elongated capsule up to 3cm long. It is not an easy plant to spot, particularly in dull weather when the flowers fold up like miniature umbrellas. The common name refers to the oval fruits, which are like tiny, shiny mirrors. The stiff hairy leaves have wavy edges and grow alternately up the stems.

Nodding or musk thistle [6]

Carduus nutans (Asteraceae or Compositae)

Biennial. Height 80–150cm. **Habitat** Grassland, wasteland, bare soil and roadsides, usually on chalky soil. **Distribution** All northern Europe as far north as south-eastern Norway. **Flowers** May–September. **Image size** Half life-size.

The slightly nodding flower heads give this thistle one of its common names, but it is better known as the musk thistle for its distinctly musky fragrance. The impressive flower heads, as much as 3–5cm across, are bright reddish-purple and consist of a mass of florets with a ruff of recurved spiny bracts below. They are usually solitary, and appear at the tip of winged stems that are armed with long spines. Together with the spiny lobed leaves, the effect is handsome enough to have been used to decorate Denby pottery. A white down on the leaves was once used for making paper. The seeds have individual parachutes of fine hairs.

Greater knapweed [1]

Centaurea scabiosa (Asteraceae or Compositae)

Perennial. Height Up to 1.2m. **Habitat** Grassland and cliffs. **Distribution** All northern Europe except Iceland. **Flowers** July–September. **In gardens** Well-drained soil in full sun. Wild-flower meadows and gardens. **Image size** Two-thirds life-size.

Greater knapweed is a magnet for butterflies and bees, which are tempted by the flamboyant purple flower heads, up to 5cm across. Each consists of a mass of florets, the outer being longer than the inner. Underneath are numerous overlapping green bracts with black fringes. The whole flower head is so hard that the plant's many other common names include hardhead and ironhead. Each leaf consists of several segments with toothed edges. The greater knapweed was once put to a whole host of medicinal uses, from curing loss of appetite to relieving catarrh and sore throats.

Rock sea lavender [2]

Limonium binervosum (Plumbaginaceae)

Perennial. Height Up to 30cm. **Habitat** Sand dunes, salt marshes and cliffs. **Distribution** Britain, France and Ireland. **Flowers** July–September. **Image size** Two-thirds life-size.

After a painstakingly slow process, the tiny flower buds of the rock sea lavender eventually unfurl to reveal papery lilac or violet-blue flowers held in attractive branched heads on upright stems. The lanceolate leathery leaves have one to three prominent veins – hence the Latin species name *binervosum* (two-veined). *Limonium* comes from the Greek *leimon*, and refers to the salt-marsh fringes where rock sea lavender puts down its roots in shingle – together with thrift and other sea lavenders (pp.127 and 163) along some coasts. The blooms are long-lasting, and cultivated plants are sold as cut-flower 'everlastings'.

Pale or striped toadflax [3]

Linaria repens (Scrophulariaceae)

Perennial. Height Up to 80cm. **Habitat** Dry habitats, waste and stony ground, and walls. **Distribution** Belgium, France and Germany; naturalised in Britain and Ireland, Holland and Scandinavia. **Flowers** June–September. **Image size** Two-thirds life-size.

Pale toadflax is rarer than common (or yellow) toadflax (p.107) with smaller flowers that vary from white to pale lilac with violet veins. Nectar is stored in the spur, which is the same length (up to 5mm) as the flower stalks. Bees are the only insects hefty enough to gain access by pressing down on the 'palate' at the mouth of the flower tube. The fine linear foliage resembles flax leaves, but the plant was considered to be fit only for a toad's use. Nevertheless, the flower shape inspired country names such as weasel snout, squeeze jaw and lion's mouth. Pale toadflax often hybridises with common toadflax to produce *Linaria* x *sepium*, which has yellow flowers with red veins.

Clustered bellflower [4]

Campanula glomerata (Campanulaceae)

Perennial. Height Up to 45cm. **Habitat** Grassland, open woodland and scrub. **Distribution** All northern Europe except Ireland, Faeroe Is and Iceland. **Flowers** June–August. **In gardens** Moist, well-drained soil in sun or partial shade. Wild-flower and woodland gardens and borders. **Image size** Two-thirds life-size.

Deep purple-blue trumpet flowers, 15–25mm long, and blood-red stems have earned the clustered bellflower the rather melancholy name Dane's blood. The upright, stiff, hairy stems emerge from a clump of rounded or heart-shaped basal leaves, which contrast with the elongated foliage growing alternately up the stems. The flowers, which can vary from violet to pale blue and even white, linger until the end of summer and are a food source for bees and butterflies. This is a vigorous species, rapidly forming substantial clumps.

Meadow crane's-bill [5]

Geranium pratense (Geraniaceae)

Perennial. Height Up to 1m. **Habitat** Meadows, woods and roadsides. **Distribution** All northern Europe except Faeroe Is and Iceland. **Flowers** June–September. **In gardens** Well-drained soil in sun. Herbaceous and mixed borders. **Image size** Two-thirds life-size.

To gardeners the meadow crane's-bill is a stalwart valued for both its deeply lobed foliage and its cup-shaped flowers. Yet it is a true native that graces meadows throughout most of Europe. The flowers – violet, blue or sometimes white – are pollinated by bees guided by darker coloured veins to the nectar. After pollination come long-beaked – crane's-bill-shaped – seedpods. When the seeds are ripe, these capsules spring apart to release them. The species has given rise to a host of garden cultivars, including the lovely 'Mrs Kendall Clark', which has pearl-grey flowers flushed rose pink.

Chicory [6]

Cichorium intybus (Asteraceae or Compositae)

Perennial. Height Up to 1m. **Habitat** Grassland, wasteland and roadsides. **Distribution** All northern Europe, but not always native. **Flowers** July–October. **In gardens** Well-drained soil in full sun. Wild-flower and kitchen gardens. **Image size** Two-thirds life-size.

Chicory – once called succory – is best known as a salad vegetable and for the coffee substitute made from its roasted roots, but it is also an attractive wild flower. The 25–40mm flowers are a clear sky-blue (or sometimes white or pink) and consist of strap-like rays with slightly toothed tips; they grow in dense clusters in the leaf axils, and make a stunning display, but open only in sunshine. The leaves at the base of the branched stems, with their triangular lobes, resemble dandelion leaves; farther up they are stalkless and clasp the stem. They have been cultivated since the 16th century, often forced and blanched by keeping them in the dark so they taste less bitter.

Harebell; in Scotland, bluebell [1]

Campanula rotundifolia agg. (Campanulaceae)

Perennial. Height Up to 50cm. **Habitat** Dry grassland, banks, heaths and dunes. **Distribution** All northern Europe. **Flowers** July–September. **In gardens** Well-drained soil in full sun or partial shade. Rock and wild-flower gardens. **Image size** Half life-size.

The apparent fragility of the harebell – with its fine upright stems and papery, nodding blue flowers – is all front. This dainty member of the bellflower family is as tough as old boots and will thrive, spreading on underground runners, in a variety of habitats; it will even doggedly push its way through tarmac. The flowers, 12–20mm long, have five reflexed lobes and stand upright in bud; they dangle only as they open. The pendent seed capsules release their seeds as they move in the wind. It is a very variable plant. The species name *rotundifolia* refers to the light green rounded or heart-shaped toothed foliage at the base. The leaves along the stem are narrower or linear, with no teeth. The flowers were said to be goblins' thimbles, and the hare of the common name is a witch's animal.

Wall speedwell [2]

Veronica arvensis (Scrophulariaceae)

Annual. Height Up to 30cm. **Habitat** Heaths, banks and walls. **Distribution** All northern Europe except Faeroe Is. **Flowers** March–October. **Image size** Half life-size.

A heathland plant by nature, wall speedwell has taken advantage of human intervention and found its ideal habitat in wall crevices, where it embeds its fine roots. The blue flowers are similar to those of the germander speedwell (p.168), with unequal petals, but on a smaller scale; they are only 3mm across and have little or no flower stalk. They grow in spikes, which account for almost a third of the upright stems. After pollination by bees, the flowers are followed by heart-shaped seed capsules. The rest of the stem has alternate oval leaves with toothed margins.

Field scabious [3]

Knautia arvensis (Dipsacaceae)

Perennial or biennial. Height Up to 1m. **Habitat** Grassy habitats. **Distribution** All except far northern Europe. **Flowers** July–September. **In gardens** Well-drained soil in sun. Borders, and wild-flower gardens and meadows. **Image size** Half life-size.

The lilac-blue flowers of field scabious are manna to pollinating bees. Tightly sealed buds open to reveal a flower head of tiny lilac or violet-blue flowers with four lobes and two prominent stamens; they stand upright when mature, so the plant looks like a collection of miniature pincushions. (The similar-looking flowers of the genus *Scabiosa*, p.185, have five lobes.) Like the true scabious, the flowers on the outer edge of the flower head of field scabious are significantly larger than the inner. The wiry stems are upright, hairy and covered in purple spots. The spear-shaped leaves become more divided farther up the stem, which they clasp; this is unlike those at the base, which have short stalks. The genus name *Knautia* is derived from the 17th-century German botanist, Dr Knaut.

Blue water speedwell [4]

Veronica anagallis-aquatica (Scrophulariaceae)

Perennial. Height Up to 50cm. **Habitat** Pond and stream margins, marshes and wet meadows. **Distribution** All except far northern Europe. **Flowers** June–August. **Image size** Half life-size.

Like the related brooklime (p.171) the water (or blue water) speedwell grows in watery places, and both are considerably more robust than the speedwells that colonise dry habitats such as roadsides. Water speedwell has thick, fleshy upright stems, which are clasped by pairs of pale green, oval leaves. Spikes of blue flowers with violet veins emerge from the leaf axils; they have the four petals and prominent stamens characteristic of the genus. It can hybridise with the pink water speedwell, *Veronica catenata*.

Hedgerow or Pyrenean crane's-bill [5]

Geranium pyrenaicum (Geraniaceae)

Perennial. Height 40–60cm. **Habitat** Grassland, hedgerows, and waste and cultivated land. **Distribution** Britain, Ireland and France; naturalised elsewhere as far north as southern Sweden. **Flowers** June–August. **In gardens** Well-drained soil in full sun or partial shade. Wild-flower gardens, and herbaceous and mixed borders. **Image size** One-third life-size.

The hedgerow crane's-bill forms clumps of hairy, rounded, evergreen foliage, with scalloped edges. Out of these clumps emerge spindly, branched stems, also covered in hairs, with purple to violet-pink or white flowers about 15mm across. The five petals are deeply notched and covered in a tracery of dark veins. Like all the crane's-bills, the flowers are followed by long, beak-like seed heads. Long-flowering and easy to maintain, it is good value for garden money, self-seeding with ease. For a pure white form, choose *G. pyrenaicum* f. *albiflorum*; the cultivar 'Bill Wallis' has deep purple flowers.

Sea aster [6]

Aster tripolium (Asteraceae or Compositae)

Annual or short-lived perennial. Height Up to 45cm. **Habitat** Salt marshes and rocky coasts. **Distribution** All northern Europe except far north. **Flowers** July–October. **Image size** Half life-size.

Fleshy leaves and stems enable the sea aster to cope with the salty environs of its coastal home, whether tidal salt marshes or the drier reaches of rocky clifftops. The daisy-like flowers consist of bright blue or purple ray florets with a mass of yellow disc florets at their centre. The leaves are linear and arranged alternately, with the flower stems arising from their axils. As long ago as the 16th century, the sea aster was a popular garden plant. However, when the Michaelmas daisy (p.206) was introduced from the New World its flamboyance soon charmed gardeners into using it in place of the more subtle sea aster.

Marsh woundwort [1]

Stachys palustris (Lamiaceae or Labiatae)

Creeping perennial. Height 10–60cm. **Habitat** Damp habitats, often near rivers, canals or ponds, and damp fields. **Distribution** Throughout northern Europe except Iceland. **Flowers** June–October. **Image size** Half life-size.

Dense spikes of pinkish-purple flowers, above very hairy toothed leaves, emerge on hairy, quadrangular stems from fleshy creeping stolons. Marsh woundwort has a faint aroma and is pollinated by bees, but the hairy corollas of the flowers bar entrance to any unwelcome insects such as flies. The roots are edible when boiled, while the young shoots can be eaten like asparagus. Among its all-healing properties, which have been exploited since the 16th century, marsh woundwort is used in modern herbal medicine as an anti-spasmodic and antiseptic; the bruised leaves are said to stop bleeding and heal wounds. It often hybridises in the wild with the darker-flowered hedge woundwort (p.176).

Opium poppy [2]

Papaver somniferum (Papaveraceae)

Annual. Height Up to 90cm. **Habitat** Cultivated and wasteland. **Distribution** Naturalised throughout northern Europe except far north. **Flowers** June–August. **In gardens** Well-drained soil in full sun. Annual and mixed borders. **Image size** Half life-size.

Originating in south-eastern Europe and Asia, the opium poppy has become a popular garden plant despite its narcotic associations; wild populations tend to be garden escapes. The large bowl-shaped flowers – up to 18cm across – grow on upright, hairy stems; they are purple, mauve, lilac or white, often with distinctive dark markings at the base of the petals. They are followed by stout blue-green seedpods; these are more rounded than those of other poppies and used in dried-flower arrangements. Opium – the source of morphine, heroin and codeine – comes from the white latex in the walls of the pods, which is extracted before they have ripened; plants growing in cool climates yield very little.

The seeds themselves are non-narcotic and edible, and are used in baking pastries and bread. 'Peony Flowered' is a frilly garden cultivar with large double flowers in shades of purple, red, pink or white.

Goat's rue [3]

Galega officinalis (Fabaceae or Leguminosae)

Perennial. Height Up to 1.5m. **Habitat** Wasteland and grassy places. **Distribution** France and Germany; naturalised elsewhere (an escapee from gardens and fields) except Faeroe Is and Iceland. **Flowers** June–July. **In gardens** Moist soil, full sun, partial shade. Borders. **Image size** One-third life-size.

Also known as French lilac, goat's rue is a native of southern and eastern Europe which was originally grown elsewhere as an ornamental plant. It is also one of several members of the large pea family grown as a fodder crop. It soon escaped, and can rampage over waste areas, providing a fine show of flower. It is a bushy, upright plant, with leaves consisting of up to 14 pairs of untoothed leaflets plus a single leaflet at the tip. Upright spikes of lilac, pink or white pea flowers, 10–15mm long, grow from the leaf axils. The 2–5cm seedpods are constricted between each seed.

Water violet [4]

Hottonia palustris (Primulaceae)

Aquatic perennial. Height Up to 40cm above water level. **Habitat** Freshwater lakes, shallow ponds and ditches. **Distribution** All northern Europe except Ireland, Iceland, Norway and Finland. **Flowers** May–July. **In gardens** Water up to 15cm deep in full sun. Wildlife and ornamental pools. **Image size** Life-size.

Statuesque and supremely elegant, the water violet throws up tall flower stems high above the water's surface, with pretty whorls of pale lilac flowers. Each has five notched petals and a pale yellow centre. The submerged foliage is thread-like, consisting of pinnate leaves that hang beneath the surface like a diaphanous underskirt; other names for the plant, such as featherfoil and millefolium, refer to this. Apart from its obvious beauty, water violet is a good plant for oxygenating water. So, where it is scarcely seen in the wild, it is widely grown in garden pools. The buds overwinter by sinking to the bottom until spring, when they rise up to the surface to produce their new growth.

Monkshood [5]

Aconitum napellus (Ranunculaceae)

Perennial. Height Up to 1.5m. **Habitat** Damp woodland, meadows, wasteland and stream margins. **Distribution** Parts of Britain, Ireland, Belgium, France and Germany. **Flowers** June–September. **In gardens** Moist soil in partial shade. Woodland gardens, and herbaceous and mixed borders. **Image size** Half life-size.

Imposing and rather grand, monkshood is aptly named for its deep purple, hooded flowers. A mass of these flowers is held in dense racemes on sturdy upright stems. The distinctive hood is formed by the sepals rather than petals. The petals have adapted to become two special nectar-containing vessels called nectaries; these are shaped like a hammer and are tucked underneath the hood. The rounded leaves are deeply lobed, with five to seven lobes which are, in turn, deeply cut.

Monkshood's attractive flowers belie the fact that it is extremely poisonous. In the Middle Ages it was cultivated to produce poison for arrows. The genus name *Aconitum* comes from the Greek *akontion* (dart); *napellus* means 'little turnip' and refers to the shape of the roots. Anyone with children should be wary of growing it in the garden; there have been cases of poisoning as recently as the 1990s.

Lesser burdock [1]
Arctium minus (Asteraceae or Compositae)

Biennial. Height Up to 2m. **Habitat** Woodland clearings, hedgerows, roadsides and wasteland. **Distribution** All northern Europe. **Flowers** July–September. **Image size** Life-size.

As tall as greater burdock (see below) and with similar-sized leaves, lesser burdock differs in the size and shape of its reddish-purple flower heads – which are egg-shaped and up to only 18mm across. The hooked bracts are tinged with purple and serve the same purposes as in greater burdock. In Scotland, since at least the 17th century, the burs have played an intrinsic part in the annual ritual of the Burry Man. This curious spectacle involves a man dressing head to toe in burs and parading the streets all day with little more than whisky to sustain him.

Greater burdock [2]
Arctium lappa (Asteraceae or Compositae)

Biennial. Height Up to 2m. **Habitat** Woodland clearings, hedgerows, roadsides and wasteland. **Distribution** All northern Europe except Faeroe Is and Iceland. **Flowers** July–September. **Image size** One-twelfth life-size.

Greater burdock has been immortalised in the foreground of many a landscape painting, the broad leaves used to fill in gaps and add perspective. The plant's stout, rounded flower heads, up to 25mm across, consist of a mass of reddish-purple florets and green, impressively hooked bracts. The species name *lappa* means 'bur', and the hooks attach themselves to clothing and animal fur – not only aiding seed dispersal but providing endless fun for children who throw them at each other. The flower heads grow in clusters on reddish, branched stems. The heart-shaped leaves, up to 50cm long, have reddish veins and are arranged in spirals up the stems.

All burdocks have a range of medicinal uses. The leaves were used to relieve sores,

while the juice soothed burns and was an antidote to snakebite. Greater burdock has several culinary uses: the roots are roasted and the young shoots peeled and eaten raw.

Wood burdock [3]
Arctium nemorosum (Asteraceae or Compositae)

Biennial. Height Up to 2m. **Habitat** Woodland clearings and disturbed ground. **Distribution** All northern Europe. **Flowers** July–September. **Image size** Half life-size.

The wood burdock is, more often than not, confused with the lesser burdock (see left). It grows in similar habitats, is the same height, and has very similar foliage. The flower heads, however, are more rounded, a darker reddish-purple and almost stalkless.

Black horehound [4]
Ballota nigra (Lamiaceae or Labiatae)

Perennial. Height Up to 1m. **Habitat** Roadsides, hedge-banks and waste ground. **Distribution** Most of northern Europe except Iceland, Faeroe Is and most of Scandinavia; possibly introduced. **Flowers** June–August. **Image size** One-and-a-half life-size.

Black horehound's very unpleasant smell when bruised defends it against being eaten by cattle, and has earned it the vernacular name 'stinking Roger'. It is a bushy plant, with oval toothed leaves in opposite pairs. The flowers grow in dense whorls; each has a funnel-shaped calyx of fused sepals, with five oval teeth, and fused purple petals forming two lips, the lower marked with white splashes.

According to William Meyrick's Herbal of 1790, black horehound is a neglected English plant 'although always possessed of great virtues'. He recommends an infusion of the leaves for 'hypochondriacal and hysterical complaints'.

Bitter vetch, vetchling [5]
Lathyrus linifolius, *L. montanus* (Fabaceae or Leguminosae)

Creeping perennial. Height Up to 40cm. **Habitat** Grassy habitats, hedgerows and scrub. **Distribution** All northern Europe except Faeroe Is and Iceland. **Flowers** April–July. **Image size** Half life-size.

The delicate sweet-pea-like 10–16mm flowers of bitter vetch start off crimson, then fade to blue or green as they mature. They are held in racemes at the tips of branched, upright stems. The slender foliage consists of two to four pairs of linear leaflets; unlike those of the sweet pea and other relatives, it ends in a point rather than a tendril. Bitter vetch is best known for its tubers, which were cultivated for food as early as the Middle Ages. They were also dried and chewed as a substitute for wild liquorice (p.229), and in Scotland sometimes used to flavour whisky.

Cotton thistle [6]
Onopordum acanthium (Asteraceae or Compositae)

Biennial. Height Up to 3m. **Habitat** Waste and cultivated land, and roadsides. **Distribution** Holland and Germany southwards; introduced in Britain and southern Scandinavia. **Flowers** July–September. **In gardens** Well-drained soil in full sun. Wild-flower and bee gardens, gravel gardens and borders. **Image size** Two-thirds life-size.

Despite being known in England as the Scotch thistle, this is almost certainly not the thistle of Scottish heraldry, which is probably the spear thistle (p.202). Yet the cotton thistle – known in Britain since the Iron Age – is an imposing plant with a spiny, winged stem and oval, wavy-edged spiny leaves. The handsome flower heads, up to 5cm across, consist of a mass of pale purple florets above a collar of cottony bracts with spiny tips. Among its myriad uses, 'cotton' fibres from the leaves and stem were used to stuff pillows and cushions, while oil from the seeds was burned as lamp fuel. It will also attract bees and butterflies into the garden and will happily self-seed.

Heather, ling [1]

Calluna vulgaris (Ericaceae)

Sub-shrub. Height Up to 60cm. **Habitat** Woodland, bogs, moors, heaths and roadsides on acid soils. **Distribution** All northern Europe. **Flowers** July–September. **In gardens** Acid soil in full sun. Ground cover, wildlife gardens and mixed borders. **Image size** Life-size.

Heather or ling forms masses of cushion-like purple mounds, and has become synonymous with heaths and moors, especially in Scotland. Yet its true natural habitats are woodland and bogs. Flowers like tiny pale purple, red, pink or white bells are suspended in racemes along slender stems that have dark green scale-like leaves arranged in pairs along their length. Named from the Greek *kallunei* (to beautify), heather has countless domestic and culinary uses. This name refers not to its decorative properties but to the use of the stems as broom bristles. The Scots also made the most of its springy properties by using it as a bedding material. It is not only a staple foodstuff for sheep but for over 4,000 years has given a honey-like flavour to a Pictish ale called Traquair. The species has given rise to more than 500 garden cultivars, many with golden foliage, some silver-grey. Planted in the garden, heather is guaranteed to attract nectar-seeking bees.

Autumnal squill [2]

Scilla autumnalis (Liliaceae)

Bulbous perennial. Height Up to 50cm. **Habitat** Rocky outcrops, grassland and coastal habitats. **Distribution** Southern England and France. **Flowers** August–October. **Image size** Life-size.

Most squills flower in spring, but this species waits until autumn, jumping the gun to produce flower stems of reddish-purple, star-like flowers with contrasting black anthers. The flowers appear long before the sparse grassy foliage. The genus name *Scilla* comes from the Greek *skilla* meaning sea onion.

Marsh gentian [3]

Gentiana pneumonanthe (Gentianaceae)

Perennial. Height Up to 60cm. **Habitat** Marshy habitats and wet acid heaths. **Distribution** Most of northern Europe, but rare. **Flowers** July–October. **Image size** Life-size.

Gentius, King of Illyria (180–167BC), is credited with discovering the medicinal properties – as an antidote to poison – of the exquisite gentians. Also desired for its shimmering blue trumpet flowers, up to 45mm long, this species is distinguished from its cousins by having five green stripes on the outside of the petals and narrow, upright foliage. Marsh gentian is a rare sight today, its numbers severely reduced by drainage and land reclamation schemes.

Large thyme [4]

Thymus pulegioides (Lamiaceae or Labiatae)

Sub-shrub. Height 5–25cm. **Habitat** Dry grassland. **Distribution** All northern Europe except Iceland and far north. **Flowers** June–September. **In gardens** Well-drained soil in full sun. Borders and rock gardens. **Image size** Life-size.

Large thyme has more elongated flowering stems than wild thyme (p.182), producing tight clusters of tiny pinkish-purple flowers, each with four lobes. However, it lacks wild thyme's self-rooting runners, forming a taller, bushier plant. The four-sided stems are angled, with hairs on the angles only – unlike wild thyme, which has hairs on two opposite sides of the stem. The slightly hairy oval leaves are arranged in pairs. The whole plant exudes a very strong aroma, and despite thyme's many culinary uses, a sprig of fresh thyme is said to bring bad luck (in particular, death or illness) into the house. On the plus side, thyme tea was used to alleviate stomach and bronchial ailments. In the garden, plant the cultivar *T. pulegioides* 'Goldentime' for its yellow foliage. Cut the stems back after flowering to maintain the plant's compact habit.

Common field speedwell [5]

Veronica persica (Scrophulariaceae)

Annual. Height 10–60cm. **Habitat** Waste, cultivated and other disturbed land. **Distribution** Naturalised in all except far northern Europe. **Flowers** All year. **Image size** Life-size.

The presence of the hairy and sprawling common field speedwell is a sure sign of human activity. An introduction from western Asia (including Persia, or Iran; hence the species name *persica*), it was first recorded in Britain in 1826. It habitually colonises disturbed tracts of waste and cultivated land, and is the most common of all the 180 *Veronica* species in northern Europe. Each vibrant blue 8–12mm flower grows on a slender stalk that arises from the axil of a hairy, toothed leaf. It has four petals, the lower one being the smallest and palest in colour. The leaves are mostly arranged alternately along the slim, gangly stems.

Corn mint [6]

Mentha arvensis (Lamiaceae or Labiatae)

Creeping perennial. Height Up to 60cm. **Habitat** Damp fields, open woodland and ditches. **Distribution** All except far northern Europe. **Flowers** July–September. **Image size** Life-size.

The tiny flowers of corn mint may be lilac or white, or occasionally pink. Like water mint (p.202), the flowers have conspicuously protruding stamens and appear in whorls as if threaded on the square stems typical of the mint family. Each whorl sits at the axil of a pair of blunt-toothed and downy leaves, which have a smell so strong and distinct that Geoffrey Grigson in *The Englishman's Flora* likened it to 'wet, mouldy Gorgonzola'. A kinder 18th-century description was of 'mellow apples and gingerbread'. Certainly the smell can ruin the quality of peppermint oil if it takes root in a crop of peppermint. Its only positive practical attribute was to discourage mice from Irish corn stacks.

Field forget-me-not [1]
Myosotis arvensis (Boraginaceae)

Annual or biennial. Height Up to 40cm. **Habitat** Dry fields, cultivated and waste ground, and dunes. **Distribution** All northern Europe. **Flowers** April–October. **In gardens** Well-drained soil in full sun. Wild-flower meadows and mixed borders. **Image size** Two-thirds life-size.

The field forget-me-not is as much loved for its delicate pale blue flowers as it is for the legend surrounding the name (p.168). The flowers, only 3–4mm across, consist of five rounded petals with a yellow eye, and grow in cymes, which are coiled when the flowers are in bud and unfurl as they open and mature. The leaves and stems are extremely hairy. At the base of the stem the linear foliage appears in a rosette; along the stems the leaves are stalkless and arranged alternately. Most annual garden forget-me-nots are derived from the larger-flowered *M. arvensis* var. *sylvatica*.

Oyster plant [2]
Mertensia maritima (Boraginaceae)

Mat-forming perennial. Height Up to 60cm. **Habitat** Coastal shingle and sandy habitats. **Distribution** Britain, Ireland, France and Scandinavia. **Flowers** June–August. **In gardens** Sharply drained, sandy soil in sun. Gravel borders. **Image size** One-and-a-half life-size.

The oyster plant is named for its attractive blue-green, spoon-shaped leaves, which have a strong oyster-like flavour from their high salt content. The foliage, which is fleshy and covered in a white bloom, is well equipped to cope with harsh, drying coastal conditions. The bell-shaped flowers, held in branched cymes, have five rounded petals; they start life as tightly furled pink buds but once open, they turn blue. Unlike other members of the family, oyster plant is completely hairless. When it was abundant, the leaves were gathered and eaten cooked or raw, but today it is getting rarer – for no obvious reason. It is available as a garden plant from specialist nurseries.

Wood vetch [3]
Vicia sylvatica (Fabaceae or Leguminosae)

Clambering perennial. Height Up to 2m. **Habitat** Rocky woodland, woodland margins, scrub and occasionally coastal habitats. **Distribution** All northern Europe except Belgium, Holland and Iceland. **Flowers** June–August. **Image size** Half life-size.

Similar to tufted vetch (p.186), the pretty wood vetch is another clambering species that uses tendrils at the leaf tips to support itself. The flowers are typically pea-like; pale lilac or white, they are shot through with dark purple veins and have an alluring scent like that of sweet peas. They are held in one-sided racemes, are fewer than on tufted vetch, and are followed by dramatic black seedpods. The leaves consist of between 5 and 12 pairs of oval leaflets. Wood vetch is one of the scarcer vetch species, and has declined along with traditional woodland management practices such as coppicing. It does thrive on sunny south-facing banks and slopes.

Sea holly [4]
Eryngium maritimum (Apiaceae or Umbelliferae)

Perennial. Height Up to 60cm. **Habitat** Coastal sandy, shingle and some rocky habitats. **Distribution** All northern Europe as far north as southern Scandinavia. **Flowers** June–September. **In gardens** Dry soil, full sun. Dry borders, gravel gardens. **Image size** One-quarter life-size.

Sea holly is instantly recognisable by its bluish-green, holly-like spiny leaves with their tough, waxy cuticle that looks like a coating of ice – arming it against harsh coastal conditions. The attractive round heads of powder-blue flowers are further shielded by substantial spiny bracts. The flowers and foliage have made it a popular architectural plant in gardens with sandy, dry soil, although it is now increasingly scarce in the wild. *Eryngium* is from the Greek *eruggarein*, to eructate or to raise gas from the stomach, implying it was a cure for indigestion and wind.

Pale flax [5]
Linum bienne (Linaceae)

Biennial or perennial. Height Up to 60cm. **Habitat** Dry grassy habitats. **Distribution** France and central and southern Britain and Ireland. **Flowers** May–September. **Image size** One-fifth life-size.

The upright, wiry stems of pale flax carry the prettiest, palest blue or bluish-lilac flowers on long stalks. Each flower has five rounded petals, but they are short-lived and tend to drop almost as soon as they have opened. The leaves are narrow and lanceolate, and are arranged alternately up the stems. Surprisingly, unlike many wild flowers growing in grassland habitats – which are always under threat – pale flax numbers are increasing.

The similar but larger-flowered common flax (*Linum usitatissimum*), an annual, is the species grown for many centuries to make linen cloth and linseed oil. The oil is still used in paints and varnishes, and the seeds are an important ingredient in cattle cake. It is sometimes found as a naturalised wild flower.

Ivy-leaved bellflower [6]
Wahlenbergia hederacea (Campanulaceae)

Trailing perennial. Height Stems up to 30cm long. **Habitat** Damp acid habitats – heaths, moors and woods. **Distribution** Britain, Ireland, Belgium, France and Germany. **Flowers** July–August. **Image size** Life-size.

The graceful ivy-leaved bellflower, with its miniature pale blue bells, trails its way through the rushes and sedges of heaths and moors. The small, pale green leaves are truly ivy-shaped, as the common and species names suggest (*Hedera* is the scientific name for ivy). They grow alternately, on long leaf stalks. The flowers appear at the leaf axils on long, spindly stems. Ivy-leaved bellflower is closely related to the genus Campanula. This genus was named for the Swedish botanist Georg Wahlenberg (1780–1851).

Spear thistle [1]

Cirsium vulgare (Asteraceae or Compositae)

Biennial. Height Up to 1m. **Habitat** Grassland, wasteland, roadsides and disturbed ground, especially on rich, fertile soil. **Distribution** All except parts of far northern Europe. **Flowers** July–October. **Image size** Two-thirds life-size.

In its youth, spear thistle forms low-growing rosettes of foliage with sharp spikes – which anyone who has had the misfortune to sit on will recognise. When about a year old, the plant produces hairy, winged flower stems up to a metre high, which carry large heads of feathery purple-red florets. Peel away the flowers and you will find a nutty core at the base of the flower head that can be eaten raw. The spear thistle is a leading contender for the role of Scotland's heraldic emblem (see also cotton thistle; p.197). Yet its reputation for being invasive and harmful to animals has given it the dubious honour of being listed as a noxious weed in the British Weeds Act of 1959.

Saw-wort [2]

Serratula tinctoria (Asteraceae or Compositae)

Perennial. Height 10–60cm. **Habitat** Grassland, heaths, scrubland, cliffs and woodland fringes. **Distribution** All except far northern Europe. **Flowers** June–August. **Image size** Two-thirds life-size.

Both the Latin and common names of this thistle-like plant refer to the toothed leaf edges, which the 16th-century herbalist John Gerard charmingly described as 'somewhat snipt around the edges like a sawe'. The thistle-like 15–20mm flower heads grow on wiry stems, and are usually purple or occasionally white; they are pollinated by bees and flies. In Germany, a greenish-yellow dye was produced from the leaves and used to tint wool (hence the species name *tinctoria*). As a herbal preparation, the whole plant was thought to mend ruptures and wounds.

Creeping thistle [3]

Cirsium arvense (Asteraceae or Compositae)

Spreading perennial. Height 0.6–1.2m. **Habitat** Pastures, meadows, roadsides and open woodland. **Distribution** All northern Europe. **Flowers** June–September. **Image size** Two-thirds life-size.

This is the most common northern European thistle. It has creeping roots that allow it to spread quickly once established. The leaves are spiny but, unlike most of its relatives, the creeping thistle produces stems that are neither spiny nor winged. The pale purple flowers are slightly fragrant and pollinated by a variety of insects. Like the spear thistle (see left), it is listed in the British Weeds Act of 1959 – meaning that it must be removed from cultivated land.

Wild teasel [4]

Dipsacus fullonum (Dipsacaceae)

Biennial. Height 1.5–2m. **Habitat** Grassland, roadsides, embankments, hedgerows, woodland margins and river-banks, usually on clay soils. **Distribution** Britain, Ireland, France, Belgium, Holland, Germany and Denmark. **Flowers** July–August. **In gardens** Well-drained soil, full sun. Wildlife gardens. **Image size** Two-thirds life-size.

A majestic, architectural plant, wild teasel in its second year produces stiff, prickly stems topped by spiky, conical 3–8cm flower heads in a delicate frame of spiny bracts. The florets are pinkish-purple to lilac. Since Roman times, the bristly dried heads were used to 'tease' out fibres from raw wool and to raise the nap of spun fabric – 'fulling'; hence the name *fullonum*. Steel combs generally replaced teasels in the 19th century, but teasels are still used for the finest cloth.

Teasel's prickly leaves are arranged in pairs and joined together at the base, making an effective natural water butt, which traps unfortunate insects. The plant may extract nutrients from the insects, while in some areas the strained water was used as a cosmetic.

Water mint [5]

Mentha aquatica (Lamiaceae or Labiatae)

Rhizomatous semi-aquatic perennial. Height 40–90cm. **Habitat** Damp woods, wet meadows, rivers and marshland. **Distribution** All except far northern Europe. **Flowers** July–September. **In gardens** Moist soil or water up to 15cm deep in full sun. Wildlife pool margins and bog gardens. **Image size** Two-thirds life-size.

More or less hairy and often purple-flushed, water mint has its leaves arranged in whorls along red-tinged stems, which are square in cross-section – a typical feature of the mint family. The toothed, oval leaves exude a powerful scent which made them popular for strewing in banqueting halls and as a type of smelling salt. Lilac, tubular flowers, 4–6mm long, appear at the stem tips in dense, rounded clusters; the mass of protruding stamens give the appearance of miniature powder puffs. For a garden pool, plant water mint in a container to keep the spreading rhizome under control, before submerging it in shallow water.

Butterfly bush, buddleia [6]

Buddleja davidii (Buddlejaceae)

Shrub. Height Up to 3m; spread up to 5m. **Habitat** Waste ground, building sites and railway embankments. **Distribution** Naturalised in all northern Europe. **Flowers** June–October. **In gardens** Fertile soil, sun or partial shade. Mixed borders, wildlife gardens. **Image size** Two-thirds life-size.

An introduction from China, this lanky bush has shown a particular flair for colonising industrial ground throughout Europe. It has long, dense spikes of honey-scented lilac flowers in summer, and grey-green lanceolate leaves. Butterfly bush is a favourite of most butterfly and moth species, which are attracted in droves when the blooms open, providing vital supplies of nectar. The flowers occasionally appear as purple or white variations. Buddleia is easy to grow and is a must for wildlife gardens, but it needs regular pruning to keep it tidy.

Alpine forget-me-not [1]
Myosotis alpestris (Boraginaceae)

Perennial. Height Up to 25cm. **Habitat** Mountain slopes, damp woods and meadows. **Distribution** Northern Britain (rare), France and Germany. **Flowers** May–September. **Image size** One-third life-size.

Short stocky stems and a covering of hairs on the stems and leaves enables the alpine forget-me-not to tolerate the high altitudes – 700 to 2,800m – of its mountainous habitats. The flowers, which start off as pinkish buds, open bright blue and are arranged in short cymes; they are larger and deeper blue than those of the otherwise rather similar wood forget-me-not (p.168). The lanceolate to oval leaves are arranged in a rosette at the base, and further up the stem are un-stalked. Alpine forget-me-not is found on the same mountain ledges as alpine fleabane (see below).

Alpine fleabane [2]
Erigeron borealis (Asteraceae or Compositae)

Perennial. Height Up to 20cm. **Habitat** Mountain ledges on chalk or limestone, and high-altitude woods. **Distribution** Scotland and Scandinavia. **Flowers** July–August. **Image size** Life-size.

It is rare to spot an alpine fleabane unless you happen to be at least 1,500m up a mountain. A short hairy plant, it produces small daisy-like flowers with lilac ray florets and yellow disc florets at their centre. The narrow leaves are lanceolate and arranged alternately. Alpine fleabane is under threat in Britain, but is less so in continental Europe, where it is found as far north as arctic Russia. In Scandinavia, alpine fleabane occurs at higher altitudes in mountain birch woodland.

Snow gentian [3]
Gentiana nivalis (Gentianaceae)

Annual. Height Up to 15cm. **Habitat** Rocky habitats, mountain meadows and marshes. **Distribution** Northern Europe except Ireland, southern Britain, Belgium, Holland and Denmark; very rare in Britain. **Flowers** June–August. **Image size** Life-size.

For the snow gentian to open its flowers at all, the light must be very bright indeed. But when they do open, the deep, intense blue of the tiny trumpet-shaped blooms is revealed. The flowers have five pointed lobes and a long narrow tube. Their stems arise from a basal rosette of oval leaves, while the stem leaves are arranged alternately. This is one of the rarest mountain plants in Britain, occurring only above altitudes of 600m in the mountains of Perth and Angus. It is under threat in Britain but, happily, thriving in continental far northern Europe.

Small toadflax [4]
Chaenorhinum minus (Scrophulariaceae)

Annual. Height Up to 25cm. **Habitat** Arable and waste land, and railway tracks. **Distribution** All northern Europe except Faeroe Is and Iceland. **Flowers** May–October. **Image size** One-and-a-half life-size.

This diminutive relation of the snapdragon (p.159) is not classified as a true toadflax (*Linaria* species; pp.177 & 190) because of the tiny pale purple flowers. They have an upper and lower lip, with a short spur behind; however, unlike true toadflax flowers – which appear to have a closed 'mouth' – the lips are parted. There is a yellow patch on the lower lip. Small toadflax produces upright, branched, hairy stems with alternate lance-shaped leaves. This is one plant that seems to be coping with the loss of its natural arable habitats by taking up residence along railway lines.

Dwarf or stemless thistle [5]
Cirsium acaule (syn. *Carduus acaulos*; Asteraceae or Compositae)

Perennial. Height 5–10cm. **Habitat** Short grassland. **Distribution** All northern Europe except Ireland, Scotland and northern Scandinavia. **Flowers** June–September. **Image size** Two-thirds life-size.

The ground-hugging rosette of the dwarf thistle's spiny leaves will be all too familiar if you have ever sat on them. It sometimes goes by the nickname 'picnic thistle'. Up to four 2–4cm flower heads appear at the centre of each rosette, consisting of purple florets above a mass of overlapping bracts. The shiny, dark green leaves are oblong in overall shape, deeply lobed and well armed with long spines on their edges.

Dwarf mallow [6]
Malva neglecta (Malvaceae)

Annual or short-lived perennial. Height Up to 50cm. **Habitat** Fields, roadsides and waste ground. **Distribution** All northern Europe except Faeroe Is and Iceland; more common in south. **Flowers** June–September. **Image size** One-quarter life-size.

Much more sprawling and weedy-looking than either common or musk mallow (p.140), this plant looks as though it has tried to be spectacular but failed. It has long-stalked kidney-shaped leaves, arising from straggling hairy stems. The pale lilac to whitish flowers, with lilac or purple veins, are up to 2.5cm in diameter and appear in clusters of three to six. Dwarf mallow seems always to be self-pollinated, producing the typical mallow 'cheese' fruits.

Water purslane [7]
Lythrum portula (Lythraceae)

Creeping annual. Height Up to 25cm. **Habitat** In and near shallow water and flooded ground; also muddy woodland on acid soil. **Distribution** All northern Europe. **Flowers** June–October. **Image size** Life-size.

Water purslane creeps on reddish stems, which send down roots from the leaf nodes, forming tangled mats of spoon-shaped foliage. The leaves are arranged in opposite pairs. Pairs of tiny purple flowers, each only 2mm across, grow at each leaf axil; they have six petals but no stalk.

Spiked speedwell [1]

Veronica spicata (Scrophulariaceae)

Perennial. Height Up to 60cm. **Habitat** Grassland, rocky habitats and wood margins. **Distribution** Britain, France and Germany; rare. Naturalised in some other areas. **Flowers** July–October. **In gardens** Well-drained soil in full sun. Rock gardens. **Image size** One-and-half life-size.

Indiscriminate farming practices, which have destroyed most of its natural habitats, have resulted in spiked speedwell being confined to only a few wild locations in northern Europe. It is now more likely to be found in domestic gardens, where it is planted for its stunning spikes of bright violet-blue starry flowers. Each 4–8mm flower has four pointed petals and protruding purple stamens. The narrow-lanceolate to oval leaves, with toothed edges, are arranged in opposite pairs up the length of the stem. Spiked speedwell attracts plenty of pollinating insects to gardens. Several cultivars are available, including 'Icicle', with white flowers.

Thyme-leaved speedwell [2]

Veronica serpyllifolia (Scrophulariaceae)

Creeping perennial. Height Stems up to 30cm long. **Habitat** Grassland, heaths, open woodland and waste ground. **Distribution** All northern Europe. **Flowers** March–October. **In gardens** Often a lawn weed. **Image size** Twice life-size.

This is a small, variable plant that roots as it creeps. The form known from gardens has upright flower spikes, with up to 30 flowers, each 6–8mm across. They have a four-lobed calyx and four petals, which are pale blue or white with lilac veins; at the base are small leaf-like bracts. In mountainous regions, the plant has a completely prostrate habit and slightly larger blue flowers, but botanists consider it no more than a subspecies. In both forms the leaves are small, oval, usually untoothed, hairless, and either stalkless or with a very short stalk. The fruit is small and heart-shaped.

Betony [3]

Stachys officinalis (Lamiaceae or Labiatae)

Perennial. Height Up to 75cm. **Habitat** Dry grassland, heaths, hedgerows and open woodland. **Distribution** All northern Europe as far north as southern Scandinavia. **Flowers** June–October. **In gardens** Well-drained soil in full sun or partial shade. Wildlife gardens, and mixed and herbaceous borders. **Image size** Half life-size.

Closely related to hedge woundwort (p.176), betony was thought to have similar healing properties when applied to wounds, earning it a place in physic gardens. However, betony was also associated with magic and it was planted in churchyards to keep evil spirits at bay. Like hedge woundwort, betony flowers have two lips, although the lower lip lacks the white markings of its cousin. The flowers are reddish-purple and occur in dense whorls that appear more oblong than round. The name *Stachys*, from the Greek meaning 'ear of corn', refers to the shape of the flower spike. Unlike hedge woundwort, the oblong leaves have few hairs but are wrinkled, deeply veined and appear in a rosette at the base of the upright stems.

Autumn gentian, felwort [4]

Gentianella amarella (Gentianaceae)

Annual or biennial. Height Up to 30cm. **Habitat** Dry hill pastures, cliffs and dunes. **Distribution** Mainly western parts of northern Europe except Iceland, northern Britain and Holland. **Flowers** June–October. **Image size** Half life-size.

Autumn gentian is an upright plant with stiff branched stems. It is placed in a separate genus from true gentians because of differences in its flower structure, including the fringe of white hairs at the top of the throat. The flowers are dull purple, blue, pink or white, with five lobes and a long tube encased in a lobed calyx. They grow in clusters on stems that emerge at the leaf axils. The oval leaves are arranged in a rosette at the base of the plant, with opposite pairs of leaves up the length of the stems.

Michaelmas daisy [5]

Aster novi-belgii (Asteraceae or Compositae)

Perennial. Height Up to 1.5m. **Habitat** Damp wasteland and river-banks. **Distribution** Widely naturalised throughout northern Europe except far north. **Flowers** September–October. **In gardens** Moist soil in full sun or partial shade. Herbaceous and mixed beds and borders. **Image size** Two-thirds life-size.

The Michaelmas daisy (named for its flowering season) arrived in Europe from North America in the early 18th century. It was so popular that it managed to oust the sea aster (p.193) to take pride of place in gardens. It continues to be successful as a garden plant and there are now huge numbers of cultivars, some with double flowers, in a wide range of colours from white through to crimson. The original species, however, produces clusters of 25–40mm flower heads with purple ray florets and yellow disc florets at their centre. The oval leaves are slightly toothed and, arranged alternately, clasp the attractive dark purple stems. Garden escapees thrive on waste ground, especially in urban areas.

Bell heather [6]

Erica cinerea (Ericaceae)

Dwarf shrub. Height Up to 60cm. **Habitat** Heathland, moorland and open coniferous woodland on dry, acid soils. **Distribution** Most of northern Europe except Iceland and far northern and eastern parts. **Flowers** July–September. **In gardens** Well-drained lime-free soil in full sun. Ground cover, peat beds and mixed borders. **Image size** Life-size.

Bell heather is aptly named for its reddish-purple miniature bell-like flowers, 5–7mm long, which hang in clusters or racemes. The 'bell' consists of four petals fused together. The evergreen leaves are dark green, smooth and needle-like, growing only 4–7mm long; fine hairs edge the rolled-back leaf margins. They are arranged in whorls of three. Like most heathers, bell heather is a magnet for pollinating bees.

Lords and ladies, cuckoopint [1]
Arum maculatum (Araceae)

Perennial. Height Up to 25cm. **Habitat** Woodland, scrub and hedgerows. **Distribution** Britain, Ireland, Holland and Germany southwards. **Flowers** April–May. **Image size** Life-size.

Lords and ladies' curious flowers have led it to be known as Adam and Eve or bulls and cows. Dutch, French, German and other colloquial names refer to the male organ – as does the 'pint' (from *pintle*) of cuckoopint. The phallic flower head in question, the spadix, is cloaked in a hood-like yellowish-green spathe – a modified leaf that is often edged and spotted with purple. Under a ring of hairs towards the spadix's base are both female and male flowers. The spadix smells bad enough to attract pollinating flies, which are trapped by the hairs until they wither. After pollination, shiny green berries – which ripen through yellow to bright red – develop on the spadix. The attractive, long-stalked, arrow-shaped leaves have dark purple spots, but all parts of the plant are toxic.

Stinking hellebore [2]
Helleborus foetidus (Ranunculaceae)

Evergreen perennial. Height Up to 80cm. **Habitat** Woods and scrub on chalky soil. **Distribution** Britain, Holland and central Germany southwards. **Flowers** January–May. **In gardens** Fertile, humus-rich soil in full sun or dappled shade. Mixed borders and woodland gardens. **Image size** Two-thirds life-size.

Stinking hellebore is aptly named. The nose-wrinkling smell of the flowers attracts early pollinating bees. The crushed leaves smell even worse, and it is no surprise that the whole plant is poisonous. Yet it is strangely attractive. Pendulous stems carry loose clusters of neat greenish-yellow cup-shaped flowers, the overlapping petals outlined in reddish-purple. The dark green, leathery leaves have 7 to 13 toothed lobes. The plant has evolved an unusual way of

dispersing its seeds. They exude an oil that attracts snails; once a snail has consumed the oil, the seed sticks to its slime and is carried away to a new site. Stinking hellebore was once used to rid children of worms, but often killed the patient too.

Dog's mercury [3]
Mercurialis perennis (Euphorbiaceae)

Creeping perennial. Height Up to 40cm. **Habitat** Woods, hedgerows and other shady habitats. **Distribution** All northern Europe except Faeroe Is and Iceland. **Flowers** February–April. **Image size** One-third life-size.

Dog's mercury spreads on creeping rhizomes, carpeting the deciduous woodland floor where it thrives. It is virulently poisonous and exudes a strong smell to attract pollinating midges. The stems are stiff, upright and unbranched; they are covered in fine hairs and contain a watery sap. The dark green oval leaves, in opposite pairs, have toothed edges. Dainty flower spikes emerge from the leaf axils, with the greenish male and female flowers on separate plants.

Butcher's broom [4]
Ruscus aculeatus (Liliaceae)

Perennial. Height Up to 75cm. **Habitat** Woods, hedgerows and rocky habitats on dry chalky soil. **Distribution** Britain and France; naturalised elsewhere. **Flowers** January–April. **In gardens** Most soils in partial or full shade. Woodland gardens, dry shady borders. **Image size** One-third life-size.

Butcher's broom is a colourful curiosity. What look for all the world like evergreen leaves are flattened stems called cladodes. They are dark green and oval, with a tough leathery texture and a spiny tip. Any true leaves are simply scales, and soon drop. The tiny flowers are dull green with purple spots. They emerge singly or in pairs on the surface of each cladode, encased in small bracts. Males and females appear on separate plants. Pollinated female plants

produce striking, glossy red berries, but be warned, they can cause stomach upsets.

Toothwort [5]
Lathraea squamaria (Scrophulariaceae or Orobanchaceae)

Parasitic perennial. Height Up to 30cm. **Habitat** Tree and shrub roots (especially alder, ash, beech, elm and hazel) in woods and hedgerows. **Distribution** Most of northern Europe except Faeroe Is and Iceland. **Flowers** March–May. **Image size** Half life-size.

Toothwort has a ghostly and rather sinister paleness about it. It contains no green chlorophyll and is a parasite that gets all the nutrients it needs from other trees and shrubs. The flowers may be creamy or pale pink, and consist of two lips with tooth-shaped scales beneath, which are actually modified leaves. They grow on short stalks in a one-sided spike. The stout fleshy stems have a mass of scales at the base.

Mistletoe [6]
Viscum album (Loranthaceae)

Semi-parasitic evergreen shrub. Spread Up to 2m across. **Habitat** Deciduous trees, including poplar, lime, hawthorn and apple. **Distribution** Most of northern Europe except Ireland, Iceland and Finland. **Flowers** February–April. **Image size** One-quarter life-size.

This is one of over 900 species of mistletoe. After birds have eaten the white berries, which appear between September and January, they deposit the sticky seeds on to a branch. There they germinate, and the roots penetrate the tree to obtain nutrients. Mistletoe is a *semi*-parasite because it produces some of its own food by photosynthesis. The woody stems fork repeatedly and produce greenish-yellow, paddle-shaped leaves, which are ribbed and leathery. The foliage is arranged in distinct opposite pairs, with a flower head of three minuscule green flowers at each axil.

Moschatel, townhall clock [1]

Adoxa moschatellina (Adoxaceae)

Rhizomatous perennial. Height Up to 15cm. **Habitat** Shady habitats on damp soil. **Distribution** All northern Europe except Holland and far north. **Flowers** April–May. **Image size** Two-thirds life-size.

Carpets of delicate moschatel spread across the woodland floor by creeping rhizomes. The alternative name, townhall clock, perfectly describes the flower heads – each has five small, greenish flowers held in tight cube-like clusters on upright stems, with four flowers facing out and one up. The outward-facing flowers have three sepals and five petals; the upward-facing one has only two sepals and four petals. The leaves are held on long stalks and consist of three lobed leaflets. The genus name *Adoxa* is a Greek word meaning 'without glory', suggesting that the plant had no culinary or medicinal significance. The species name *moschatellina* describes its musky scent.

Bath asparagus [2]

Ornithogalum pyrenaicum (Liliaceae)

Bulbous perennial. Height Up to 80cm. **Habitat** Woods, scrubland and meadows. **Distribution** Southern Britain, Belgium and France. **Flowers** April–June. **In gardens** Well-drained soil in full sun. Borders and woodland gardens. **Image size** One-quarter life-size.

Bath asparagus owes its name to the Avon valley, where it occurs in great abundance around Bristol and Bath, and to the fact that the unopened flower spikes were traditionally picked in May and sold at market for cooking and eating like asparagus. The starry, greenish-white six-petalled flowers appear in a spiked raceme and have a very faint fragrance. Bluish-green, strap-like leaves, similar to those of the bluebell (p.170), emerge, mature and die long before the flower buds open.

Wild asparagus [3]

Asparagus officinalis subsp. *prostratus* (Liliaceae)

Prostrate perennial. Height Stems up to 1.2m long. **Habitat** Grassland, scrub and wasteland. **Distribution** Northern Europe as far north as Denmark and southern Sweden. **Flowers** June–August. **Image size** Two-thirds life-size.

Unlike the familiar garden asparagus (*A. officinalis* subsp. *officinalis*, an introduction from southern Europe that is also – rarely – found in the wild), wild asparagus is a northern European native. But it also is rare, often thriving best near coasts. The stems are prostrate, with what appear to be clusters of between four and 15 needle-like leaves; but these are actually adapted stems called cladodes. The tiny flowers are greenish-white and bell-shaped, with male and female flowers on separate plants. Orange berries follow the female flowers. The juicy shoots of both wild and naturalised garden asparagus are edible.

Herb paris [4]

Paris quadrifolia (Liliaceae)

Perennial. Height Up to 40cm. **Habitat** Moist woods on chalky soil. **Distribution** All northern Europe except Ireland and far north. **Flowers** May–June. **In gardens** Moist, leafy soil in full or partial shade. Woodland gardens. **Image size** Life-size.

Herb paris is an eccentric plant whose name actually means 'herb of pairs'. Each stem has, halfway up, a whorl of four broad oval leaves with a solitary flower arising from the centre. Each starry flower has four spear-shaped sepals between four slender, greenish petals; there is a crown of eight yellowish stamens in the centre. A fleshy black fruit follows, which splits open to release its seeds. The consistent numbers of the plant's parts – which in reality vary – gave it an aura of magic, and it was reputed to ward off the evil effects of witchcraft. Yet the berries are poisonous enough to cause stomach upsets if eaten.

Sheep's sorrel [5]

Rumex acetosella (Polygonaceae)

Perennial. Height Up to 30cm. **Habitat** Short grassland, bare ground and open heaths on acid, sandy soil. **Distribution** All northern Europe. **Flowers** May–August. **Image size** Half life-size.

Sheep's sorrel's inconspicuous greenish flowers are wind-pollinated, so they have no need for flamboyant colours to attract insects. Instead, the male flowers, which are on separate plants from the females, produce copious amounts of pollen. Both are borne in branched spikes. The sparse leaves are arrow-shaped, with two lobes at the base; they point forwards or spread outwards and in late summer turn a bright crimson. It is notoriously difficult to tell many sorrels apart, although the fruits are a reliable means of identification; in sheep's sorrel they are small and, unlike several other species, have no warts. Like all sorrels, this species contains calcium oxalate, which gives it a sour taste; the juice is poisonous if consumed in quantity.

Common sorrel [6]

Rumex acetosa (Polygonaceae)

Perennial. Height Up to 1m. **Habitat** Grassland, roadsides, riverbanks and woodland. **Distribution** All northern Europe. **Flowers** May–June. **In gardens** Rich, fertile soil in full sun or partial shade. Herb and vegetable gardens. **Image size** Half life-size.

For centuries, sour-tasting common sorrel leaves have been used in fish dishes and salads. As in sheep's sorrel (see above), the sourness is strongest at the height of the season – early on it is almost tasteless. Common sorrel throws up grooved, stems with branched flower spikes at their tip. Like sheep's sorrel, the greenish male and female flowers appear on separate plants. The arrow-shaped foliage has two back-pointing lobes at the base of each leaf; the upper leaves clasp the stem. The flowers, fruit and foliage all turn crimson later in the season.

Early spider orchid [1]

Ophrys sphegodes (Orchidaceae)

Perennial. Height Up to 20cm. **Habitat** Grassy places and scrub on chalky soil. **Distribution** South-eastern Britain, Belgium and Germany southwards. **Flowers** April–June. **Image size** Three times life-size.

The pale to dark brown lip of the early spider orchid looks more like a bee than a spider. The contrasting petals and sepals are a pale yellowish-green and the oval lip carries a bluish-violet X or H marking. The oval leaves are arranged at the base of the flower spike. The early spider orchid at best sets only a little seed, possibly accounting for the fact that it is rare. The closely related late spider orchid (*Ophrys fuciflora*; syn. *O. holoserica*) has a broader brown lip with the violet-blue X or H marking outlined in yellow; the petals and sepals are pink. It flowers in June and July. Its habitats and range are similar and it is also rare.

Fly orchid [2]

Ophrys insectifera (syn. *O. muscifera*; Orchidaceae)

Perennial. Height 40–60cm. **Habitat** Woodland, scrub, grassland, fens and lake margins, usually in shade on chalky soil. **Distribution** All northern Europe except Faeroe Is and Iceland. **Flowers** May–June. **Image size** One-and-a-half life-size.

The flowers of the fly orchid look as if they have a fly alighting on them. But if you look carefully you will see that is, in fact, a cunning device for attracting pollinating insects, as with the closely related bee orchid (p.132). In this case, the target is the male *Gorytes* wasp. The flower is elaborately constructed from three greenish sepals and a velvety purple-brown lip, which mimics the fly-like female wasp. The lip has three lobes, the central one having a forked tip and a pale violet marking towards the base. The flowers release pheromones to attract the male wasps. The flowers are arranged in a slender spike with narrow, lanceolate leaves arising from its base.

Frog orchid [3]

Coeloglossum viride (Orchidaceae)

Perennial. Height Up to 20cm. **Habitat** Grassland, wood margins and scrub, often on hills. **Distribution** All northern Europe. **Flowers** June–September. **Image size** Two-thirds life-size.

It takes a lively imagination to spot the resemblance, but the frog orchid's flowers do look vaguely like a frog's head. They are inconspicuous, so the plant is difficult to find, particularly as it rarely grows any higher than the surrounding vegetation. The lightly scented flowers are yellowish-green with a purple or reddish tinge and consist of a hood formed by sepals and petals to protect the pollinia and stigma; a notched lip protrudes from the hood. As the flower matures, the hood opens to reveal the reproductive parts. Two to five oval leaves clasp the stem. The frog orchid is scarce in many regions but pockets of plants can be found on chalky downs and hilly pastureland.

Man orchid [4]

Aceras anthropophorum (Orchidaceae)

Perennial. Height Up to 40cm. **Habitat** Grassland and scrub on chalky soils. **Distribution** England, Holland, Belgium, France and Germany. **Flowers** May–June. **Image size** One-and-a-half life-size.

The dainty and rare man orchid is so called because the greenish-yellow flowers look vaguely like human figures. They grow in a tall, slender spike and each has a hood (the 'head'), made up of sepals and petals, and a lip. The lip has side lobes – the 'arms' – and a central lobe, which is divided into two 'legs'. The flowers have no spur (unlike related orchids of the genus *Orchis*) and are sometimes edged with red. The leaves are shiny, oblong and arranged at the base of the flower spike. The flowers are scented and the leaves contain coumarin, which gives off the pleasant scent of new-mown hay as the foliage dries.

Bird's-nest orchid [5]

Neottia nidus-avis (Orchidaceae)

Saprophytic perennial. Height Up to 50cm. **Habitat** Deciduous woodland, especially shady beech woods on chalky soil. **Habitat** All northern Europe except Faeroe Is and Iceland. **Flowers** May–July. **Image size** Half life-size.

This ghostly orchid is a saprophyte, a plant that gets all its nutrients from dead and decaying plant matter (unlike a parasite, which feeds on living plants). It does so with help from fungi called mycorrhiza, which grow around the roots, so it thrives in woods where there is plenty of leaf litter. It has no leaves and no green chorophyll, so the erect stems and flowers are yellowish-brown. Overlapping scales up the length of the stem take the place of leaves. The flowers smell sickly-sweet and have a lip with two lobes beneath the sepals and petals. The plant gets its name from the mass of tangled fleshy fibres at the base of the stem.

Common twayblade [6]

Listera ovata (Orchidaceae)

Perennial. Height 20–60cm. **Habitat** Woods, hedgerows, grassland and sometimes marshy ground. **Distribution** All northern Europe except the far north. **Flowers** May–July. **Image size** One-quarter life-size.

This is one of the most common orchids in northern Europe, including Britain, although it usually remains unnoticed. This is despite its distinctive single pair of leaves near the base, hence the name twayblade (two-blade). They are oval, with prominent veins that run from the base of each leaf to its tip. As many as 100 yellowish-green flowers are borne in an upright spike; each consists of five petals and a long, narrow, furrowed lip. They are pollinated by beetles and flies tempted by the nectar at the base of the lip, which they reach by climbing up the furrow. At the top of the furrow they brush their heads against the pollinia (modified stamens) and collect pollen, which they deposit on to the stigma of the next plant.

Common broomrape [1]

Orobanche minor (Orobanchaceae)

Parasitic perennial. Height Up to 60cm.
Habitat Meadows and other grassy habitats, and
hedgerows; mainly on clovers. **Distribution** Most of
northern Europe except Belgium, Faeroe Is and Iceland;
naturalised in Denmark, Sweden and Ireland.
Flowers June–September. **Image size** Life-size.

The parasitic common broomrape lives
mainly off the roots of members of the pea
family and if it gets a grip it can devastate a
commercial crop of clover (*Trifolium*
species). It also parasitises members of the
daisy family and other soft-stemmed plants.
The stem, which is swollen at the base, is
yellowish tinged with red or purple, and
carries slightly hairy yellowish flowers
consisting of a two-lipped corolla. The
flowers are fewer on each stem and smaller
(at 10–18mm) than those of either ivy or tall
broomrape. Like all species of broomrape,
common broomrape contains no
chlorophyll, so the scale-leaves are brownish;
they are arranged spirally around the stem.

Greater broomrape [2]

Orobanche rapum-genistae (Orobanchaceae)

Parasitic perennial. Height Up to 90cm. **Habitat** Rough
grassland; mainly on gorse and broom. **Distribution**
Most of northern Europe except Scotland, northern
Ireland, Faeroe Is, Iceland and Scandinavia. **Flowers**
May–July. **Image size** Life-size.

The name broomrape comes from this
particular species, which obtains its
nutrients from the roots of shrubby
members of the pea family, particularly
gorse and broom. The 'rape' part of the
name is from the Latin *rapa*, meaning
'turnip-like' and referring to the obviously
swollen base of the stem. Containing no
green pigment, the stems are pale yellow
with brownish scale-leaves. The flowers
are up to 25mm long, and are yellowish
with a tinge of purple; the lower lip has
a hairy margin.

Tall or knapweed broomrape [3]

Orobanche elatior (Orobanchaceae)

Parasitic perennial. Height Up to 75cm. **Habitat**
Grassland; on knapweeds and globe thistles. **Distribution**
Most of northern Europe as far north as Sweden.
Flowers June–July. **Image size** Two-thirds life-size.

Tall broomrape is, like the other
broomrapes (see left and below), a parasite.
It gets all the nutrients it needs from greater
knapweed (p.190) and some related
members of the daisy family, on whose
roots it grows. The stems are a reddish- or
yellowish-brown. The flowers are similar to
those of ivy broomrape, except that, in this
species, they are yellowish, tinged with
purple, and the upper lip is notched. The
scale-leaves are brownish in colour and
grow spirally around the stem.

Ivy broomrape [4]

Orobanche hederae (Orobanchaceae)

Parasitic perennial. Height Up to 60cm. **Habitat**
Woods, hedgerows and old walls; on ivy.
Distribution Southern England, Wales and Ireland.
Flowers June–July. **Image size** Half life-size.

The broomrapes are a family of parasites.
They contain no chlorophyll, which green
plants use to synthesise nutrients, so each
species obtains its nutrients from a
particular 'host' plant. As its name suggests,
ivy broomrape specialises in growing on
the roots of common ivy (*Hedera helix*).
In the absence of green pigment, the thick,
upright stems and the scales that take the
place of true leaves are a reddish-brown
colour. The dull cream flowers are tinged
purplish-brown and consist of a two-lipped
corolla with a narrow neck. The top lip
has two lobes and the bottom lip three.
The scale-leaves are arranged spirally
around the stem.

Sea sandwort [5]

Honckenya peploides (Caryophyllaceae)

Creeping perennial. Height 5–25cm. **Habitat** Coastal
sand, shingle and rocks. **Distribution** All northern
Europe. **Flowers** May–August. **Image size** Life-size.

Wherever you find sea sandwort, you can be
sure that it is toiling away to prevent erosion
of the shoreline. Anchored by penetrating
roots, its bright green prostrate stems are
thick with layers of fleshy foliage and root
themselves at the nodes. In this way it creeps
across sand and shingle, creating mini-
windbreaks and, in the process, lays the
foundations for new sand dunes. Sea
sandwort is usually dioecious – that is, the
6–10mm male and female flowers appear on
separate plants. They are easily distinguished
from one another: the male flowers have five
relatively obvious greenish-white petals,
while the females have such tiny petals that
at first sight they appear non-existent. The
fruits are like small green peas.

Sea beet [6]

Beta vulgaris subsp. *maritima* (Chenopodiaceae)

Sprawling perennial. Height Up to 1.5m. **Habitat**
Coastal shingle, salt-marsh edges and sometimes grassy
banks. **Distribution** Most of northern Europe as far north
as southern Sweden. **Flowers** June–September.
Image size One-quarter life-size.

The humble sea beet is the progenitor of an
important group of plants that includes
sugar beet, beetroot and leaf-beet (spinach
beet). It has been cultivated for at least
2,000 years – first of all in the Middle East
– for its long fleshy roots and nutritious
leaves. However, in the wild, sea beet is a
dull plant with a straggly, sprawling form.
The upright flowering stems are red-tinged,
and the dark green heart-shaped leaves have
a tough, leathery texture to withstand the
rigours of their seaside habitat. Dense
clusters of small green flowers are arranged
in spikes and consist of five segments, each
containing a yellow stamen.

Great burnet [1]
Sanguisorba officinalis (Rosaceae)

Perennial. Height Up to 1.2m. **Habitat** Damp grassland and woods. **Distribution** All northern Europe except parts of north. **Flowers** June–September. **In gardens** Moist, well-drained soil in sun or partial shade. Wild-flower meadows. **Image size** Two-thirds life-size.

Great burnet produces a mass of tiny brownish-red flowers, which are densely packed into an oblong flower head. The flowers have no petals, but each has four brown-red sepals and four prominent stamens. They are held aloft on bare, wiry stems, and are pollinated by insects attracted by the copious nectar. The leaves are pinnate, consisting of three to seven pairs of oval leaflets with toothed edges. Both leaves and roots were at one time used to relieve diarrhoea, dysentery and bleeding – hence *Sanguisorba*, from *sanguis* (blood) and *sorbeo* (to staunch).

Deadly nightshade [2]
Atropa belladonna (Solanaceae)

Perennial. Height Up to 2m. **Habitat** Damp shady habitats, woodland clearings and scrub, on chalky soils. **Distribution** Britain, Belgium and Germany southwards; naturalised in Ireland, Denmark and Sweden. **Flowers** June–August. **Image size** Two-thirds life-size.

Every bit as poisonous as its name suggests, deadly nightshade can kill children and adults alike. The berries, flowers, leaves and roots all contain alkaloid poisons, yet in 16th-century Venice it was used cosmetically, to dilate the pupils of the eyes – hence the species name *belladonna*, 'beautiful lady'. Three centuries later chemists discovered the active agents, which include atropine and hyoscyamine, and these are still used to dilate the pupils for eye examinations. This member of the potato family is a well-branched plant with large oval leaves. A few solitary violet-brown, bell-shaped flowers, 25–30mm long, grow from the leaf axils and they are

followed by the tempting 15–20mm shiny green berries, which later turn black. Note that bittersweet or woody nightshade (p.172) is sometimes misnamed deadly nightshade.

Great water dock [3]
Rumex hydrolapathum (Polygonaceae)

Perennial. Height Up to 2m. **Habitat** Marshes, river and canal banks, and pond and lake margins. **Distribution** Most of northern Europe except northern Scandinavia and Iceland. **Flowers** July–September. **Image size** Two-thirds life-size.

There are some 200 species of dock, and the fruits are often the only clue as to their identity. However, the appearance of great water dock is distinctive. It is a stately, handsome plant of shallow water and watersides with sturdy branched stems and huge banana-like leaves up to 60cm long. They are dull green. The tiny, insignificant greenish flowers appear in pendulous clusters and consist of six perianth segments; in great water dock they have no teeth. The fruits are triangular with an oblong pimple at their tip.

Fat hen [4]
Chenopodium album (Chenopodiaceae)

Annual. Height Up to 2m. **Habitat** Waste, cultivated and bare ground and roadsides. **Distribution** All northern Europe. **Flowers** June–October. **In gardens** A weed. **Image size** One-third life-size.

One fat hen plant can produce up to 75,000 seeds. This, along with its ability to colonise bare ground in a matter of weeks, has made it a highly successful plant competitor since the Bronze Age, when it was almost certainly eaten by humans. The leaves are the most conspicuous part, and can vary from diamond-shaped to lanceolate; they are generally toothed and have pale undersides. The insignificant, greyish-white flowers appear in spikes at the leaf nodes. More nutritious than spinach, the leaves have a similar flavour and can be added to salads. Birds are always pleased to feast on the bountiful supply of seeds.

Black bindweed [5]
Fallopia convolvulus (syns. *Bilderdykia convolvulus* and *Polygonum convolvulus*; Polygonaceae)

Twining annual. Height Stems up to 1.5m long. **Habitat** Arable and waste land, and roadsides. **Distribution** All northern Europe. **Flowers** July–October. **Image size** One-third life-size.

Seeds of black bindweed have been unearthed in several Bronze Age archaeological digs, suggesting that it was cultivated and the seeds eaten, some 3,000 years ago. The whole plant is covered in a fine down. Its thin, twining stems bear heart- or arrow-shaped leaves. Loose spikes of tiny greenish-pink or greenish-white flowers grow from the leaf axils; they have five perianth segments. The fruits – which give black bindweed its name – are dull black nutlets with three wings.

Greater plantain [6]
Plantago major (Plantaginaceae)

Perennial. Height Up to 40cm. **Habitat** Grassy places and wasteland. **Distribution** All northern Europe; naturalised in Iceland. **Flowers** June–October. **In gardens** A lawn weed. **Image size** One-third life-size.

Distinguishing greater plantain from other plantain species is straightforward, simply because it has obviously longer flower spikes that take up almost half the scape, or flower stem. The flowers are wind-pollinated. Prominent anthers allow their pollen to be easily lifted by the wind, while the stigma protrudes just enough to catch it. The foliage grows in one or more rosettes at the base of the flower stems. The supple oval, ribbed leaves can tolerate extensive trampling under foot on pathways, tracks and lawns.

Annual nettle [1]

Urtica urens (Urticaceae)

Annual. Height Up to 60cm. **Habitat** Cultivated ground and wasteland. **Distribution** All northern Europe. **Flowers** June–October. **In gardens** A weed, but attracts butterflies. **Image size** One-and-a-half life-size.

Unlike the common stinging nettle (see below), annual nettle is monoecious, meaning that it has both male and female flowers on the same plant. Tiny and greenish, the flowers appear in compact clusters in the leaf axils. The foliage is more rounded and more deeply toothed than the common species and, although it does have stinging hairs, they are not quite as vicious.

Humans have a love-hate relationship with nettles. As weeds, they tend to follow in our footsteps, colonising disturbed ground and indicating that the soil is rich in nutrients. However, the stems were woven to make fabric and the iron-rich leaves – which lose their sting when cooked – are consumed in nettle soup and nettle tea. Nettles also attract breeding butterflies and moths, whose caterpillars adore the leaves.

Common or stinging nettle [2]

Urtica dioica (Urticaceae)

Perennial. Height Up to 1.5m. **Habitat** Woodland, fens and cultivated ground, on rich soil. **Distribution** All northern Europe. **Flowers** June–September. **In gardens** A weed, but useful food plant for caterpillars. **Image size** One-and-a-half life-size.

Nettles are wind-pollinated so have no need of the flamboyant blooms that attract insects. The tiny male and female greenish-white flowers of the common stinging nettle appear on separate plants. The females grow in long pendulous clusters; clusters of male flowers stick out at right-angles from the stem in the leaf axils. Opposite pairs of heart-shaped leaves are deeply toothed and, like the stem, are armed with vicious stinging hairs. The sting is caused by the tiny hairs breaking off when touched and

releasing acid into the skin. Nettle stings were believed to protect against sorcery and the plants themselves thought to prevent milk from being soured by witches.

Curled dock [3]

Rumex crispus (Polygonaceae)

Perennial. Height Up to 1m. **Habitat** Disturbed, cultivated and waste land, hedgerows, marshes and pond margins. **Distribution** All northern Europe. **Flowers** June–October. **Image size** One-fifth life-size.

Docks are notoriously difficult to tell apart, but curled dock has distinctive, large lanceolate leaves, up to 30cm long, with obviously wavy margins. Unlike many other dock leaves, they don't have a heart-shaped base. The thick stems are upright and branched, and carry dense sprays of greenish, bisexual flowers in whorls on pendulous stalks. The fruits are small and brown with three swollen wings, or 'valves'. Docks often grow among nettles, and curled dock is one of the species (along with broad-leaved dock, *Rumex obtusifolius*) used to relieve nettle stings. The leaves contain a soothing astringent and are also used to make a herbal ointment to repair cuts and soothe itching.

Northern dock [4]

Rumex longifolius (Polygonaceae)

Perennial. Height Up to 1.2m. **Habitat** Damp grassland, ditches, and river and lake margins. **Distribution** Most of northern Europe except Holland. **Flowers** June–July. **Image size** Life-size.

Northern dock is so called because in Britain it mostly grows northwards from Lancashire and Yorkshire. This dock species has long, narrow leaves – hence the species name *longifolius* ('long-leaved') – with wavy edges. Tiny greenish flowers appear in dense spikes. The wings or valves of the fruits are kidney-shaped.

Intermediate lady's mantle [5]

Alchemilla xanthochlora (garden syn. *Alchemilla vulgaris*; Rosaceae)

Perennial. Height Up to 50cm. **Habitat** Grassy places. **Distribution** Northern Europe as far north as central Scandinavia. **Flowers** May–July. **In gardens** Moist, humus-rich soil in full sun or partial shade. Wild-flower gardens and herbaceous borders. **Image size** One-third life-size.

There are about 250 very similar species of *Alchemilla*, and their attractive foliage and sprays of yellow-green flowers have made them popular garden plants. The best known is *A. mollis* from south-eastern Europe, which has escaped to grow wild in some areas. But *A. xanthochlora* is also grown, often misnamed *A. vulgaris* by nurserymen and gardeners – a name that properly belongs to another species (p.221). It produces clumps of kidney-shaped leaves with hairy undersides and leaf stems but – unlike many of the other species – has a smooth upper leaf surface. Masses of tiny yellow-green flowers appear in cymes in early summer; *xanthochlora* is from Greek words literally meaning 'yellow-green'.

Sibbaldia [6]

Sibbaldia procumbens (Rosaceae)

Perennial. Height Up to 5cm. **Habitat** Mountain grassland and rocky slopes. **Distribution** All northern Europe except Ireland, Belgium and Holland. **Flowers** July–August. **Image size** One-and-a-half life-size.

Sibbaldia was named after Sir Robert Sibbald, professor of medicine at Edinburgh University in the 16th century. Low-growing and tufted, the plant is perfectly adapted for its hostile mountainous environment, with hairy leaves to reduce water loss. It has trifoliate leaves with oval leaflets that have three lobes or teeth at the tip. The tiny yellowish-green flowers are arranged in clusters on reddish stems and have sepals that are longer than the petals.

Good King Henry [1]

Chenopodium bonus-henricus (Chenopodiaceae)

Perennial. Height Up to 50cm. **Habitat** Cultivated, waste and bare land, roadsides and old walls. **Distribution** All northern Europe except far north. **Flowers** May–August. **Image size** Two-thirds life-size.

Good King Henry is a glamorous name for an unassuming plant. It comes from the German *Guter Heinrich* – from a character in folklore – not to be mistaken for 'Bad Henry', the poisonous dog's mercury (p.209). Its stiff, upright stems carry spirals of triangular leaves; the juvenile foliage is covered in white powder or meal. The small yellowish-green flowers sometimes have a red tint and appear in spikes. Good King Henry has colonised wasteland for centuries; despite this, the young leaves and flowering tips were boiled and eaten with butter.

Common lady's mantle [2]

Alchemilla vulgaris (Rosaceae)

Perennial. Height Up to 60cm. **Habitat** Damp grassy and rocky habitats, roadsides, woodland margins and streamsides. **Distribution** All northern Europe except far north. **Flowers** June–September. **Image size** Two-thirds life-size.

The true *Alchemilla vulgaris* – as distinct from the garden plant often wrongly given the same name (p.218) – has velvety foliage encrusted with jewel-like drops of moisture on humid days. These are not dew or raindrops but water forced out of tiny pores by the lobed leaves. The droplets were collected by herbalists, who believed that they would heal wounds and cure infertility – hence the name *Alchemilla* ('little alchemist'). Later the plant was named for the Virgin Mary – in Norway it is called *Marikaabe* (Mary's mantle). The British name translates the German *Frauenmantel*; 'mantle' refers to the cloak-like leaves. Wiry flower stems push their way through these clumps of foliage, carrying clusters of tiny greenish-yellow flowers.

Petty spurge [3]

Euphorbia peplus (Euphorbiaceae)

Annual. Height Up to 30cm. **Habitat** Cultivated and waste ground. **Distribution** Northern Europe as far north as central Scandinavia, except Faeroe Is and Iceland. **Flowers** April–October. **Image size** Two-thirds life-size.

There is nothing petty about this spurge other than its short stature, for it is poisonous enough to make livestock very ill. It contains a potent milky sap in both the branched stems and the leaves, which are bright green and rounded. The tiny yellowish-green flowers appear in the leaf axils and consist of a cluster of male flowers surrounding a single female; the whole is encased in green bracts. The sap was once used to burn off corns and warts.

Caper spurge [4]

Euphorbia lathyris (Euphorbiaceae)

Biennial. Height Up to 2m. **Habitat** Shady habitats on bare and cultivated land. **Distribution** Britain (Sussex). Naturalised elsewhere. **Flowers** June–July. **In gardens** Well-drained, light soil in sun or shade. Woodland and wild gardens (but may become a weed). **Image size** Two-thirds life-size.

The poisonous fruits of caper spurge have occasionally been mistaken for the edible buds of the true caper – hence the common name. It is an introduction from central and southern Europe, although in a few parts of southern England it is considered a true native. It is a stiff, upright plant with a blue-green tinge to the waxy, stemless and almost oblong leaves, which grow in opposite pairs. In its second year, caper spurge produces small male and female flowers cupped by much more prominent bright green, triangular bracts. The flowers are followed by the caper-like fruits. Like all spurges, this species contains a poisonous and irritant white sap. For some, caper spurge is a garden weed, while for others it is an attractive ornamental plant worthy of cultivation.

Common figwort [5]

Scrophularia nodosa (Scrophulariaceae)

Perennial. Height Up to 1m. **Habitat** Hedgerows, woods and other damp, shady habitats. **Distribution** All northern Europe except far north. **Flowers** June–September. **Image size** One-third life-size.

Common figwort emerges from rhizomes, or underground stems, which produce small nodules. These not only gave the plant the species name *nodosa* but were thought to look like piles or glands. From this came its Medieval use to treat scrofula, a form of tuberculosis that causes swollen neck glands. From that in turn came the genus name *Scrophularia*. Common figwort's stems are upright, stiff and branched at the tips. They carry opposite pairs of oval leaves with sharply toothed edges. The purple-brown flowers consist of two lips – the upper longer than the lower – and give off an unpleasant smell that attracts wasps. Green figwort (*S. umbrosa*) is similar but has leaves that are pointed and without lobes.

Pellitory-of-the-wall [6]

Parietaria judaica (Urticaceae)

Spreading perennial. Height Up to 40cm. **Habitat** Walls, rocks, cliffs and steep hedge-banks. **Distribution** Britain, Ireland, western parts of northern Europe. **Flowers** June–October. **Image size** Half life-size.

Pellitory-of-the-wall (or wall pellitory) is a lot more benign than its close relative the stinging nettle (p.218). Instead of leaves and stems covered in stinging hairs, it has smooth oval leaves with red-tinged veins and sprawling, reddish stems. The tiny greenish-yellow flowers are arranged in clusters of three or more up the stem and positioned at the base of a leaf. In the 16th century, it was thought to cure kidney and bladder stones, because it is robust enough to push its way through stone. The common name is a corruption of its Middle English name *paritorie*, which came from the Latin *parietarius*, meaning 'belonging to the walls'.

Portland spurge [1]
Euphorbia portlandica (Euphorbiaceae)

Annual or short-lived evergreen perennial. Height Up to 40cm. **Habitat** Coastal sand dunes and short grass. **Distribution** Southern and western Britain and Ireland, France. **Flowers** July–October. **In gardens** Well-drained, light in full sun. Rock gardens. **Image size** Two-thirds life-size.

There are over 1,500 species of *Euphorbia* and in this vast genus there is a wide variety of growth habits and characteristics. Portland spurge is a bushy plant with slightly succulent evergreen foliage that has evolved to minimise moisture loss in demanding coastal conditions. The leaves have a leathery texture, a narrow lance shape and a silvery-grey hue. They have a prominent midrib, and are arranged alternately up fleshy stems that sometimes turn reddish. The male and female flowers, which appear on branched flower stems, are cupped in bright yellowish-green triangular bracts.

Sun spurge [2]
Euphorbia helioscopia (Euphorbiaceae)

Annual. Height Up to 50cm. **Habitat** Waste and cultivated land; roadsides. **Distribution** All northern Europe except far north. **Flowers** May–August. **In gardens** A weed. **Image size** One-third life-size.

Sun spurge is one of the most common *Euphorbia* species in northern Europe, but is, however, poisonous. As with most spurges, the male and female flowers are separate, but you have to inspect them very closely to find that one 'flower' consists of a single female ovary on a stalk surrounded by a cluster of tiny male flowers. The whole arrangement is held in a cup called an involucre, which itself is enveloped by rounded yellowish bracts. A number of these grow in a broad umbel, facing directly up towards the sun. (The species name *helioscopia* comes from Greek words meaning 'sun gazer'.) Like several spurges, sun spurge produces segmented seed heads that explode with a loud bang to release the

seed. Once the seeds are scattered on the ground, ants attracted by oil in a sac attached to each seed carry them off. After they've consumed the oil, they abandon the seeds to germinate.

Dwarf spurge [3]
Euphorbia exigua (Euphorbiaceae)

Annual. Height Up to 20cm. **Habitat** Arable and waste land on chalky soil. **Distribution** Most of northern Europe except Faeroe Is, Iceland and Finland. **Flowers** June–October. **Image size** Twice life-size.

Like many spurge species, dwarf spurge grows mostly on land disturbed by human activity and is a common cornfield weed. Unlike most of its relatives, this is a dainty plant – the species name *exigua* means 'very little' – with the length of the slender stems covered in narrow linear leaves. The flowers, which are actually a perianth-like structure called a cyathium, have greenish triangular bracts and kidney-shaped glands, which have distinctive long horns. The genus *Euphorbia* is named after Euphorbus, 1st-century physician to the king of Mauritania, credited with being the first to use the plant medicinally. Its uses range from burning off warts to relieving rheumatism.

Sea spurge [4]
Euphorbia paralias (Euphorbiaceae)

Perennial. Height Up to 60cm. **Habitat** Coastal sand dunes, shingle and rocks. **Distribution** Britain, Ireland, Belgium, Holland and France. **Flowers** July–October. **Image size** One-quarter life-size.

Taller than Portland spurge (see left), sea spurge is otherwise very similar and has just the same characteristics that help it to tolerate dry and salty coastal conditions. Fleshy, reddish stems are smothered in overlapping succulent grey-green leaves, but they lack Portland spurge's prominent midrib. The umbels of male and female

flowers are protected by grey-green oval bracts. Sea spurge is poisonous.

Scottish asphodel [5]
Tofieldia pusilla (Liliaceae)

Perennial. Height Up to 20cm. **Habitat** Wet, mountainous habitats. **Distribution** Most of northern Europe except Ireland, central and southern Britain, Belgium, Holland and Denmark. **Flowers** June–August. **Image size** One-third life-size.

The diminutive yet elegant Scottish asphodel produces fans of flat, sword-shaped leaves. The flower spike emerges from the centre of the foliage bearing rounded clusters of greenish or white flowers at its tip. They are followed by rounded fruits. Scottish asphodel is closely related to the bog asphodel (p.108) and thrives among damp mosses. The slightly more robust and showy Tofield's asphodel (*Tofieldia calyculata*) of central Europe is a better plant for moist garden sites.

Wild madder [6]
Rubia peregrina (Rubiaceae)

Evergreen trailing or scrambling perennial. Height Stems up to 1.5m long. **Habitat** Open woodland, hedgerows, scrub and rocky habitats. **Distribution** South-western and southern Britain, Ireland and southern and western France. **Flowers** June–August. **Image size** Two-thirds life-size.

Armed with rough prickles, the square, trailing stems of the wild madder clamber their way through brambles and other supporting plants in a rampant fashion. Dark green oval, leathery foliage grows in whorls of between four and six leaves, and the margins are armed with hooked teeth. Branched flowering stems arise from the leaf axils and carry pale yellowish-green, star-like 5mm flowers in loose panicles. The black berries that follow become fleshier as they ripen. The roots of wild madder yield a rose-pink dye.

Clustered dock [1]
Rumex conglomeratus (Polygonaceae)

Perennial. Height Up to 60cm. **Habitat** Damp grassy habitats, and river and pond margins. **Distribution** All northern Europe. **Flowers** June–October. **Image size** Two-thirds life-size.

Reddish zig-zag stems are the distinguishing feature of clustered dock among a large genus of plants that are often near-impossible to tell apart (pp.163, 217 and 218). The dark green leaves are narrow and pointed, with reddish veins. Loose clusters of reddish-brown flowers with narrowly oval petals emerge from the leaf axils. Wherever clustered dock grows, it is more than likely that winter flooding occurs on a regular basis.

Common glasswort, marsh samphire
[2] *Salicornia europaea* (Chenopodiaceae)

Annual. Height Up to 35cm. **Habitat** Salt marshes and mudflats. **Distribution** All northern Europe. **Flowers** August–September. **Image size** Two-thirds life-size.

The branched stems of glasswort, dried and burned, were used to make glass because of their high mineral content. Common glasswort is well adapted to its salty coastal environment, having succulent cactus-like stems, scale-like leaves and minuscule flowers of one stamen and three styles. The edible young stems can be eaten raw, or simmered and eaten like asparagus; or they can be pickled in vinegar. When in fruit, in September and October, the stems are flushed red or pink.

Spear-leaved orache [3]
Atriplex prostrata (syn. *A. hastata*; Chenopodiaceae)

Annual. Height Up to 1.5m. **Habitat** Waste and cultivated land in coastal habitats. **Distribution** All northern Europe except far north. **Flowers** July–October. **Image size** Two-thirds life-size.

Spear-leaved orache is a statuesque plant on sturdy square stems and can be identified by its strongly arrow-shaped leaves with their distinctive square base. (In other oraches the leaves taper into the leaf stems.) Greenish male and female flowers grow on the same plant and are arranged in dense panicles. The fruits are protected by pairs of triangular bracts armed with teeth. The name orache comes from the Greek *atraphaxis* – also the basis of the genus name *Atriplex*; but the origin of the Greek name itself is obscure.

Sea purslane [4]
Halimione portulacoides (syn. *Atriplex portulacoides*; Chenopodiaceae)

Perennial or sub-shrub. Height Up to 1m. **Habitat** Salt marshes and coastal pools. **Distribution** Northern Europe as far north as central Scotland and Denmark. **Flowers** July–October. **Image size** Two-thirds life-size.

Glittering mounds of sea purslane are a striking sight along salt-marsh fringes, but this 'daughter of the sea' is also perfectly adapted to its harsh coastal environment, which includes regular flooding. The silvery colour of the foliage is caused by a layer of minuscule scales on the leaf surfaces. These reflect the intense light and trap moist air, protecting the plant against desiccating coastal winds and salt water. The oval leaves have no teeth and are arranged in opposite

pairs. The greenish-yellow male and female flowers generally grow together on one plant in clusters, although occasionally they appear on separate plants.

Hop [5]
Humulus lupulus (Cannabaceae)

Climbing perennial. Height Stems up to 6m long. **Habitat** Hedgerows, often climbing up trees and telegraph poles, and scrub; also escaped from cultivation. **Distribution** Most of northern Europe except far north. **Flowers** July–September. **In gardens** Moist, well-drained soil in sun or partial shade. On fences and walls. **Image size** Two-thirds life-size.

The aromatic oils released by female hop flowers have been the vital ingredient of beer (as opposed to ale, made without hops) since the Middle Ages. The male and female flowers grow on separate plants and are quite distinct. The pale yellowish-green males (shown in the main photograph) are borne in lax clusters and are of no use in brewing. The tiny yellow female flowers (shown on this page) are enclosed in overlapping scales, which look rather like miniature cones. The leaves are arranged in pairs and have three to five toothed lobes. The stems are covered in tiny hook-like hairs, which support the plant as it clambers.

Frosted orache [6]
Atriplex laciniata (Chenopodiaceae)

Annual. Height Up to 30cm. **Habitat** Sandy beaches and dunes. **Distribution** All northern Europe except Faeroe Is, Iceland and Finland. **Flowers** July–September. **Image size** Half life-size.

The silvery foliage earns this species its common name; the effect is caused by a white mealy covering. The diamond-shaped leaves are deeply toothed and arranged in spirals up reddish or yellowish stems. Clusters of tiny yellowish-green flowers emerge at the leaf axils. It is a common sight on sandy beaches, where it grows alongside prickly saltwort (*Salsola kali*).

Parsley piert [1]

Aphanes arvensis (Rosaceae)

Annual. Height Up to 10cm. **Habitat** Stony soil, bare ground, wasteland and roadsides. **Distribution** All northern Europe except Faeroe Is, Iceland and most of Scandinavia. **Flowers** April–October. **Image size** Twice life-size.

Parsley piert is a ubiquitous plant of stony soil. Its common name was adopted in part because it vaguely resembles parsley, but it originally came from the French *perce-pierre* (pierce-stone). It grows among and between stones, so was also called parsley break-stone. Herbalists took this habit as a sign that it could break up kidney and bladder stones. The sprawling mound of hairy grey-green foliage consists of tiny leaves, each with three segments that are lobed at their tips. The minuscule green flowers have four sepals instead of petals and are arranged in clusters, each protected by a stipule armed with triangular teeth.

Mountain sorrel [2]

Oxyria digyna (Polygonaceae)

Perennial. Height Up to 30cm. **Habitat** Damp rocky places and by streams, mainly in mountains. **Distribution** Most of northern Europe except Denmark and Holland. **Flowers** June–August. **Image size** Life-size.

Unlike the arrow-shaped leaves of many related sorrel species, mountain sorrel has attractive kidney-shaped foliage. This grows on long reddish stems from the base of the plant and the leaves take on pretty reddish hues in autumn. The flower stems are stiff, upright and branched; they carry spikes of minute greenish flowers with four petals. The flowers are followed by nutlets, which droop on short pendulous stems and have wings that equip them effectively for wind distribution. Even though it is related to the edible common sorrel (p.210), mountain sorrel has none of common sorrel's culinary attributes.

Bog orchid [3]

Hammarbya paludosa (syn. *Malaxis paludosa*; Orchidaceae)

Bulbous perennial. Height 8cm. **Habitat** Sphagnum bogs. **Distribution** All northern Europe. **Flowers** July–September. **Image size** One-and-a-half life-size.

The elusive bog orchid is not only rare but very difficult to find because of its diminutive proportions and erratic flowering. Added to that, it grows in the uninviting environs of sulphurous-smelling bogs. When it does bloom, slim upright stems are smothered in tiny greenish-yellow flowers of five petals and a lip. They differ from most orchid flowers because they are upside-down, with the lip at the top of the flower. The tips of the sparse oval leaves have a row of tiny bulbils, like tiny beads, which detach and fall to the ground, so the plant can propagate itself even without producing seed. Instead of roots the bog orchid has fine hairs, which absorb nutrients from the rich peat bog. It is a protected species in Northern Ireland.

Dutchman's pipe, yellow bird's-nest

[4] *Monotropa hypopitys* (Monotropaceae or Pyrolaceae)

Saprophytic perennial. Height Up to 30cm. **Habitat** Damp beech and pine woodland. **Distribution** All northern Europe except Faeroe Is and Iceland. **Flowers** June–September. **Image size** Half life-size.

The ethereal Dutchman's pipe is a saprophyte of anaemic appearance. It lacks the green pigment chlorophyll, so the whole plant, including the scaly leaves, is creamy-white or yellowish, turning brown as it ages. Instead of manufacturing food by photo-synthesis, it absorbs nutrients from decaying leaf litter via a 'nest' of fleshy roots at the base of the stem that houses threads of decaying fungi. With its colouring and nodding cluster of bell-like flowers, it looks like a clay pipe – hence the common name. The alternative name yellow bird's-nest refers to the roots. Once the rounded fruits ripen, the flower head becomes upright.

Annual mercury [5]

Mercurialis annua (Euphorbiaceae)

Annual. Height Up to 40cm. **Habitat** Waste and cultivated land. **Distribution** All northern Europe except Faeroe Is and Iceland; naturalised in Ireland and Scandinavia. **Flowers** July–October. **In gardens** A weed. **Image size** Life-size.

Make no mistake: all parts of annual mercury, like those of the perennial dog's mercury (p.209), are dangerously poisonous. This is true even after they have been boiled, even though the leaves were once eaten as a vegetable and also used in enemas. Annual mercury looks rather similar to dog's mercury, with its stiff, upright stems and pairs of toothed oval leaves. In both species the green male and female flowers appear on separate plants. However, annual mercury is more branched and lacks the creeping rhizomes of dog's mercury, spreading instead by producing hundreds of tiny seeds. Once established it tends to stay put, so it needs ruthless eradication from gardens, especially if there are children around.

Grass-leaved orache [6]

Atriplex littoralis (Chenopodiaceae)

Annual. Height Up to 1.5m. **Habitat** Salt marshes and other coastal habitats. **Distribution** All northern. Europe except Faeroe Is and Iceland. **Flowers** July–October. **Image size** One-quarter life-size.

Grass-leaved orache has sparse linear – literally grass-like – leaves that set this species apart from its relatives. Their shape helps the plant to endure drying, salty conditions because it minimises the leaves' surface area to reduce moisture loss. The stems are striped and have a slight zig-zagged appearance. Clusters of greenish flowers are borne in long spikes, which arise from the leaf axils. Male and female flowers are found in separate spikes on the same plant; the females have no petals but two bracts. The fruits are enclosed by a pair of toothed, diamond-shaped bracts.

Broad-leaved helleborine [1]

Epipactis helleborine (Orchidaceae)

Perennial. Height Up to 80cm. **Habitat** Woods (especially beech), banks and scrub, on chalky soil. **Distribution** All northern Europe. **Flowers** July–September. **Image size** Two-thirds life-size.

Broad-leaved helleborine is an ethereal plant of beech woodland, and bears masses of flowers – sometimes up to 100 – on a single flower spike. As in most orchids the waxy, slightly nodding flowers have a complicated structure. In this species they consist of greenish sepals, petals that range from pale pinkish-violet to deep purplish-red, and a rounded lower lip. The deeply veined leaves are broadly oval – hence the common name – and grow in a spiral up the stem. The roots are infected with mycorrhizal fungi, which provide the plant with vital nutrients. Broad-leaved helleborine is uncommon; although it is essentially a woodland plant, in Glasgow it grows quite happily in parks and cemeteries, on waste ground and railway embankments, and in quarries – a phenomenon that occurs in no other British city.

Annual pearlwort [2]

Sagina apetala (Caryophyllaceae)

Annual. Height Up to 15cm. **Habitat** Dry, bare ground, sandy heaths and paths. **Distribution** All northern Europe. **Flowers** April–August. **Image size** Twice life-size.

At first sight, annual pearlwort is easily confused with the perennial procumbent species (see right). But on closer inspection it is seen to have less sprawling stems, with no basal rosette. The solitary green flowers appear on shorter stalks and despite the name *apetala*, meaning without petals, they always produce petals, although they are short-lived. The leaves are long and pointed. Pearlworts are tougher than they look and will tolerate plenty of trampling underfoot; as a result they are able to colonise well-worn paths.

Procumbent pearlwort [3]

Sagina procumbens (Caryophyllaceae)

Perennial. Height Up to 20cm. **Habitat** Paths, short grass and damp shady habitats. **Distribution** All northern Europe. **Flowers** May–September. **In gardens** A weed **Image size** Life-size.

Procumbens means 'growing along the ground', so this mat-forming, moss-like species is aptly named. Sprawling stems spread from a tufted central leaf rosette. The linear leaves, which have bristly tips, appear to be arranged in whorls but are actually in closely-spaced pairs. Tiny green, solitary flowers emerge on long, slender stems; some have four rounded petals, but others have none at all. Procumbent pearlwort has hallowed associations; it was supposedly the first plant upon which Christ stepped on Earth, when rising from the dead. Disappointingly, the pretty name pearlwort has no basis in romantic folklore; it was simply supplied by botanists, who may have been referring to the unopened flowers, which are pearl-shaped.

Wood sage [4]

Teucrium scorodonia (Lamiaceae or Labiatae)

Perennial. Height 30–50cm. **Habitat** Open woodland, scrub, hedgerows, shingle and dunes. **Distribution** All northern Europe as far north as southern Norway. **Flowers** July–September. **Image size** Half life-size.

Wood sage is one of the few members of the mint family to have little fragrance. It smells more like hops than anything else, and indeed has been used in brewing, giving beer a bitter flavour. Nevertheless, it is an elegant plant with slender, upright stems and spikes of dainty pale greenish-yellow flowers, 8–9mm long, which have a conspicuous lower lip and contrasting brownish stamens. The sage-like leaves are heart-shaped with toothed edges and are arranged in opposite pairs. The genus name comes from Teucer, ancestor to the kings of Troy, who used the plant medicinally.

Wild liquorice [5]

Astragalus glycyphyllos (Fabaceae or Leguminosae)

Perennial. Height 80–150cm. **Habitat** Grassland, scrub and open woodland, mostly on chalky soils. **Distribution** Most of northern Europe except Ireland, Faeroe Is, Iceland and far north. **Flowers** June–August. **Image size** Life-size.

Despite the species name *glycyphyllos*, meaning 'sweet leaf', it is not wild liquorice but a related southern European plant, *Glycyrrhiza glabra*, that is used to make liquorice confectionery. In fact, wild liquorice is a type of milk-vetch; the genus name *Astragalus* refers to the belief that it increased goats' milk yield. Its stout, straggly stems have a zig-zagged appearance, and the foliage is typical of vetches, with oval leaflets arranged in opposite pairs. The tubular flowers are held in dense racemes; their greenish-cream colour sets it apart from other vetches and milk vetches. The flowers are followed by long curved seedpods, which are divided lengthwise.

Marsh cudweed [6]

Gnaphalium uliginosum (syn. *Filaginella uliginosa*; Asteraceae or Compositae)

Annual. Height Up to 25cm. **Habitat** Wet and damp places, especially bare ground. **Distribution** All northern Europe. **Flowers** July–October. **Image size** Twice life-size.

Marsh cudweed is an insignificant little grey plant, which lurks in puddles and rutted paths and by ponds. The rounded, sprawling stems and alternate, lanceolate leaves are smothered in silver-grey woolly hairs. At the tip of the stems, up to 10 dirty yellow-brown flowers are arranged in a cluster and surrounded by narrow leaves. Heath cudweed (*G. uliginosum*) produces flowers in spikes rather than clusters. The genus name comes from a Greek word for 'wool', and the soft leaves, covered in woolly hairs, were used to stuff cushions. Cudweed had some use as a gargle, while the leaves, bruised with fat, were once fed to cows that had lost the ability to chew their cud.

An overview of wild-flower conservation in Britain and Ireland

Corncockle (*Agrostemma githago*, page 152) was once a common weed of cornfields, so abundant that it was considered a menace because its poisonous black seeds spoiled the flavour of bread. Today this cornfield annual, with its pretty purple flowers that unfurl like a miniature flag, are all but extinct in the wild. Until the 1950s, the vivid blue spikes of the perennial meadow clary (*Salvia pratensis*, page 173) made a striking spectacle in open grassland, along sunny hedge-banks and on woodland margins. Populations of this handsome yet now elusive plant have dwindled to only a few sites in the south of England. Similarly the delightful Deptford pink (*Dianthus armeria*, page 139) with its dainty pink flowers and blue-grey leaves once a common coloniser of tracks, dry pastureland and hedgerows throughout England, now inhabits only 7 per cent of its original area. It is extinct in Scotland. The conversion of pasture fields to arable and forestry, and the increase in building development on pastureland has taken its toll.

Over 200 species of wild flower, including Deptford pink and meadow clary, are now so threatened that they, along with over 150 species of wild flower, have been afforded legal protection by the Wildlife and Countryside Act 1981 (page 232). In Ireland, the Wildlife Act of 2000 protects

almost 70 wild flowers. Nevertheless, in Britain one species of wild flower is still lost from each county every two years (Plantlife 2017). Between 1960 and the late 1980s, over 20 species of wild flowers disappeared from wetland habitats alone. According to the most recent research carried out on British wild flowers, one in five are facing extinction. Yet now the greatest concern is not for those plants we already know to be rare, but those that we could previously take for granted as being widespread. Wild flowers such as corn buttercup (*Ranunculus arvensis*) and dodder (*Cuscuta epithymum*, page 147) are on the decline and now classified as endangered or vulnerable in *The Vascular Plant Red Data List for Great Britain* of 2005. Other parts of northern Europe tell a similar story, for example, in Germany approximately 270 species of wild flowers are considered endangered

and nearly 50 species are now extinct (*www.plant-talk.org*, 2005).

Reasons for decline

Several factors are responsible for this sad decline in wild-flower populations. The major reasons are the introduction of exotic flora, habitat loss and indiscriminate plant collection. The loss of wild-flower habitats is, without doubt, the most significant factor and largely the result of intensive farming methods, urbanisation, peat and mineral extraction and pollution. In the UK, 98 per cent of wild-flower meadows, nearly 85 per cent of open heathland, 96 per cent of open peat bogs and since the Second World War nearly 50 per cent of hedgerows have disappeared. In the 21st century, airports continue to expand, motorways are widened and immense

LEFT Mixed with acorns in wine, greater stitchwort (*Stellaria holostea*) could cure a stitch, but today it is a threatened species along with its hedgerow habitat.

RIGHT An introduction from Australasia, the aquatic *Crassula helmsii* is a rampant invader of shallow wetlands, which out-competes many native plants.

Spanish invader: the vigorous Spanish bluebell (*Hyacinthoides hispanica*, page 181) should not be mistaken for the native English bluebell.

Elegant and elusive, summer lady's tresses (*Spiranthes aestivalis*) is a native of damp grassland, but habitat loss has rendered it extinct in parts of northern Europe.

housing developments are encouraged by government. As a result the destruction of wild-flower habitats is an issue that regularly comes under the spotlight. After habitat loss, the introduction of invasive plants from outside Britain and Ireland is considered to have had the second-greatest impact on native wild-flower species (Royal Horticultural Society, 2005). These plants are often ruthless competitors for nutrients, moisture and light; in establishing themselves they oust less vigorous native species. In Britain the English bluebell (*Hyacinthoides non-scripta*, page 170) is under threat from the Spanish bluebell (*Hyacinthoides hispanica*, page 181). The latter is a garden escapee, which hybridises with the English species to produce vigorous hybrid offspring. The English bluebell is unable to compete – the result is its ongoing slow decline. Other examples include floating pennywort (*Hydrocotyle ranunculoides*), which has

successfully reduced populations of aquatic plants such as frogbit (*Hydrocharis morsus-ranae*, page 68). For centuries wild plants have been collected for culinary, medicinal and economic purposes, but not least for their beauty. Some species have become a victim of their own success, for example, the lady's slipper orchid (*Cypripedium calceolus*), whose numbers were greatly reduced by indiscriminate collection. Other wild-flower populations have been eroded by unscrupulous collectors who systematically collect plants for illegal sale; bulbous plants, such as the English bluebell (page 170), are particularly vulnerable.

Legislation to conserve wild flowers and their habitats

Several measures are in place to protect wild flowers. The conservation of habitats began as far back as 1949 when the British

government introduced statutory protection for threatened areas of countryside and created 11 National Parks from as far south as Dartmoor up to Northumberland. The specific purposes of the National Parks is to 'conserv[e] the natural beauty, wildlife and cultural heritage of those areas' and 'promot[e] opportunities for public understanding and enjoyment of... the Parks'. The 1949 statute also gave rise to Areas of Outstanding Natural Beauty, National and Local Nature Reserves. These areas all play a vital role in protecting valuable wild-flower habitats, while Sites of Special Scientific Interest (SSSIs) are regularly used as a weapon in planning wars against indiscriminate development. Of these, National Nature Reserves are the only areas where wild flowers are protected for their conservation value.

The Habitats Directive is a vital European initiative, which safeguards wild flowers through the creation of protected areas across the European Union. These areas are known as Natura 2000 sites and include Special Areas of Conservation (SACs), Special Areas of Protection (SAPs) and SSSIs (see above). The *Important Plant Areas* programme is designed to ensure that these areas are given the best possible protection from existing legislation. It is linked with a number of European and global conservation programmes, such as Natura 2000 and the European Plant Conservation Strategy.

In 1981 the British government passed the Wildlife and Countryside Act, which made it a criminal offence for anyone to dig up wild flowers without the permission of the landowner. Over 150 of the most rare species of wild flowers are recorded on an endangered list and it is a criminal offence to pick them, dig them up or destroy them without a licence – selling these plants is illegal. Nine British plants are protected by the Habitats Directive from picking, uprooting, destruction and sale, including the marsh saxifrage, creeping marshwort and fen orchid. The Bern Convention, which came into force in 1982, aims to conserve wild flora and fauna in their natural habitats, with emphasis on cooperation across Europe.

The Convention on International Trade in Endangered Species (CITES) is a further international agreement between governments that acts as a measure against the illegal sale of wild-flower species. By listing endangered species of wild flower that could suffer from illegal trade, it aims 'to ensure that international trade in specimens of wild animals and plants does not threaten their survival'. The snowdrop (page 31) is the only British wild-flower species to be included on the CITES list.

Not all rare and endangered plants are protected by legislation, but the *Red List of Threatened Species* records and highlights all those wild-flower species that are threatened with extinction. Research has shown that densely populated European countries, such as Britain, which are farmed intensively and have highly developed infrastructures, are more likely to find a higher proportion of their native wild-flower species recorded on the *Red List of Threatened Species*. On the other hand, in Scandinavia, which is less densely populated, fewer native wild-flower species appear on the *Red List*.

Other conservation measures
Until now farmers and their agricultural practices have shouldered much of the responsibility for the loss of wild flowers

RIGHT The English bluebell (*Hyacinthoides non-scripta*) is under threat from the Spanish bluebell, which hybridises with the native species to produce highly competitive offspring.

Motorway embankments are just one type of development that can provide unexpected opportunities for the creation of new wild-flower habitats. Occasionally wild flowers, such as the once-scarce cowslip (*Primula veris*), will colonise these unlikely sites of their own free will.

and their habitats. Attitudes, however, are changing for the better. Most significantly, farmers across the European Union are now encouraged through reforms to the Common Agricultural Policy to manage areas of their farmland for conservation purposes. In England the Environmental Stewardship Scheme was launched in 2005 as a direct result of these reforms. The scheme, promoted by the Department for Environment, Food and Rural Affairs (DEFRA), provides farmers with the incentive to earn money for increasing biodiversity on their land by maintaining hedgerows, creating wild-flower habitats and protecting ponds. Farmers are also learning through education that areas such as field margins, field islands, tree islands and hedgerows are less profitable for crop production and better developed as environmental habitats, according to the Farmed Environment Company in 2005.

Urban conservation

Conservation of wild flowers is not exclusive to the rural environment, being just as valuable in an urban context. New initiatives, such as those implemented by the British environmental charity Landlife (see opposite), where new wild-flower habitats are created on waste or derelict city land, vastly improve what would otherwise remain neglected tracts of land. New plantings of wild flowers also provide important opportunities for maintaining and increasing populations of threatened species and their associated insects and animals, while bringing an appreciation and understanding of the importance of wild flowers to communities that would otherwise rarely or never come into contact with them.

Despite what appears to be the wholesale destruction of wild-flower habitats where development or urbanisation takes place, occasionally these provide the ideal conditions for wild flowers to re-establish themselves and important opportunities for the imaginative implementation of new wild-flower populations. It was not so long ago that the cowslip (page 84) was a rarity, yet it is now a ubiquitous plant of motorway and trunk road embankments, with most populations having established themselves where few other plant species seem to colonise. On the British side of the Channel Tunnel near Dover, chalk marl exposed by the development has been planted with wild-flower mixes that replicate populations of maritime plants already found growing in the area including rock sea lavender (*Limonium binervosum*, see page 190) and golden samphire (*Inula crithmoides*, page 112). The M3 motorway at Twyford Down in Hampshire was a controversial development, yet radical steps were taken to save populations of wild flowers that would have been destroyed by the motorway. Turf, rich in horseshoe vetch, was literally dug up and relocated, while thousands of seeds and plug plants including rock rose, kidney vetch, cowslip clustered bellflower and devil's-bit scabious were planted to create new downland wild-flower habitats replacing those that had been lost as a result of the development (*www.floralocale.org*, 2005).

Conservation bodies

There are several societies and organisations that exist to conserve and re-establish populations of wild flowers in Britain and Ireland. They also provide a means through which the public can become involved in protecting local wild-flower species.

The Botanical Society of the British Isles uses its expertise to conserve wild flowers through the publication of national atlases and county floras. It also maintains the definitive database of wild flowers in Britain. The Society runs a comprehensive education programme from A-level standard up to postgraduate level.

English Nature is funded by the Government to promote the conservation of England's wildlife and natural features. Several acts of Parliament, including the Wildlife and Countryside Act 1981 (page 232), enable English Nature to carry out important conservation measures.

Flora Locale encourages good practice among native wild-flower suppliers, land managers and those involved in large-scale habitat creation and restoration projects in the countryside. The organisation advises on a number of projects countrywide that are committed to preserving and encouraging populations of wild flowers. These include school projects, the restoration of wild-flower habitats on disused arable farmland and the recreation of habitats on areas of development such the Channel Tunnel. Flora Locale also runs a series of courses devoted to the subject.

Botanic Gardens Conservation International (BGCI) is a membership organisation representing a network of over 500 botanic gardens in more than 100 countries around the world. BGCI promotes and supports the plant conservation and environmental education activities of botanic gardens, with a specific focus on achieving the targets of the Global Strategy for Plant Conservation. In the UK, BGCI works with its members to ensure the conservation of threatened plants, both in living collections and in seedbanks.

The Ministry of Defence may be an unlikely champion of flora and fauna but it owns several sites that are managed for conservation purposes. These include Salisbury Plain in Wiltshire, which is a tank training ground but is also one of the

Landlife is an organisation devoted to creating new wild-flower habitats in the urban environment under the banner of 'creative conservation'. Working to bring the community and conservation together, it has succeeded in improving otherwise derelict land as well as developments such as motorway embankments.

largest surviving, semi-natural dry grasslands in Europe. It supports populations of the rare burnt-tip orchid, the green-winged orchid (page 178) and the frog orchid (page 213) (www.defence-estates.mod.uk, 2005).

Plantlife is a charitable organisation devoted to conserving wild flowers and habitats. Plantlife owns 22 wild-flower reserves throughout England, Wales, Scotland and the Isle of Man. The reserves cover a range of habitats including wild-flower meadows, ancient woodland, chalk grassland, blanket bog and limestone pavements. To date the organisation has rescued 30 of the UK's most rare plants from extinction through 'Back from the Brink', a conservation programme launched in 1991. The programme currently covers 101 species of wild flower including meadow clary (*Salvia pratensis*, page 173). Conservation work consists of a combination of field and laboratory research, advice to land owners on the management of wild flowers and hands-on action by Flora Guardians, a team of volunteers who carry out practical tasks such as dredging ponds, clearing scrub, monitoring populations of wild flowers and caring for the wildlife reserves. Plantlife is closely associated with the European organisation, Planta Europa (see right).

Royal Botanic Gardens, Kew carries out extremely valuable work associated with the conservation of plants. The Threatened Plants Appeal not only benefits exotic species of plants worldwide but is also in place to protect our native wild flowers. As a result of funds raised by the ongoing Appeal, conservationists have the resources 'to check the wild status of the plants, assess genetic diversity and record methods of propagation and maintenance'. The aim of the programme is to 'repatriate or reintroduce [species of wild flower] to protected areas of natural habitat…' The Millenium Seedbank at Kew's outstation at

Wakehurst Place in West Sussex is a global initiative to protect 24,000 species of plants against extinction. To date, the project has secured the future of nearly all of Britain's native flowering plants. (www.rbgkew.org.uk, 2005)

Royal Society for the Protection of Birds owns 142 nature reserves countrywide primarily for the protection of Britain's birds, but in so doing they protect and maintain valuable wild-flower habitats.

The Wildflower Society is for amateur wild-flower enthusiasts who enjoy searching for and recording species in their natural habitats. The Society was established in 1886 and, over 100 years later, its principle aim remains educating the public, and particularly the younger generation, in the art and science of field botany, the conservation of wild flowers, British flora and the countryside. The Society supplies members with a Field Botanist's Record Book, which lists 1,000 of the most common wild flowers in Britain and in which they can record sightings.

The Wildlife Trusts is a partnership of local organisations and a campaigning organisation that works with government, industry, landowners and communities to protect wildlife and wildlife habitats by raising awareness of the potential threats to wildlife. There are 47 local Wildlife Trusts across Britain, which manage 2,500 nature reserves between them. Some of their work includes the restoration of grasslands that support rare species of orchids and the snake's-head fritillary (page 167).

Ireland There are a number of wild-flower conservation initiatives including a wild-flower seed bank of Irish rare and threatened plants which, sourced from native Irish plants, serves to protect the Irish wild-flower gene pool. The seed bank is a collaboration between Trinity College Botanic Garden, the Irish National Botanic

RIGHT An imaginative planting by Landlife shows that wild flowers can succeed almost anywhere.

Garden, the National Parks and Wildlife Service and the Irish Genetic Resources Conservation Trust.

The Irish Wildlife Trust has campaigns in place to protect and conserve valuable wild-flower habitats such as wetlands – which are at risk from pollution, drainage and most recently a significant increase in development – and hedgerows which, like those in Britain, have been vastly reduced as a result of intensive farming practices.

Conservation Volunteers Ireland is an organisation that works in partnership with appropriate bodies and landowners to replant and maintain wild-flower meadows.

The National Parks and Wildlife Service safeguards Ireland's National Parks, Special Areas of Conservation, Special Protection Areas and National Heritage Areas, all vital habitats for Ireland's wild flowers.

Planta Europa is 'a network of independent organisations, non-governmental and governmental, working together to conserve European wild plants and fungi'. (*www.plantaeuropa.org, 2005*). The organisation aims to halt the loss of wild plant diversity in Europe by 2007. Planta Europa is closely associated with Plantlife International (see above).

For contact details, turn to page 248.

PLANT COLLECTION

It should be stressed that this book does not intend to encourage the collection of any wild flowers from their natural habitats.

Gardening with wild flowers

The humble bumblebee pollinates at least 80 per cent of the food crops we eat. The irony is that agriculture is, by and large, mainly responsible for endangering or causing the extinction of at least 25 per cent of the 256 species of wild bees found in Britain. Acres of meadowland and thousands of miles of hedgerows have been lost to intensive farming methods, taking with them wild flowers and nest sites crucial for sustaining bee populations (see An Overview of Wild-flower Conservation, p.231).

Happily, though, urban and rural domestic gardens are becoming increasingly important as havens for threatened native wild flowers and the wildlife that seek them out for food and nesting sites. Gardeners, too, benefit from gardening with the environment in mind. Wild flowers will not only enhance the garden with their beauty, but also attract insects that are expert at pollinating our garden fruits and flowers; some insects, along with birds and small mammals, are also indispensable predators against aphids, slugs and other garden pests.

The following gives a brief overview of some of the wild-flower habitats that gardeners can create and short lists of recommended wild flowers for each. For more detailed information, consult any of the specialist publications devoted to gardening with wild flowers and gardening for wildlife listed in the Further Reading on page 249.

LEFT Wild teasel (*Dipsacus fullonum*), moth mullein (*Verbascum blattaria*) and chicory (*Cichorium intybus*) are as beautiful growing in a garden as they are in their natural habitats and attract beneficial pollinating insects.

An instant blaze of colour from cornfield annuals, such as corncockle (*Agrostemma githago*) and corn marigold (*Chrysanthemum segetum*), can be achieved in the garden in only a few months.

Recreating wild-flower habitats

Where possible, it is important to try to introduce plants into the garden that are local to the area. This information is available from the Postcode Plants Database (page 248).

Wild flowers should always be purchased from reputable nurseries and suppliers. The plants must come from cultivated stock and not wild collected populations, which is an illegal practice.

In Britain and Ireland wild-flower seed should be of local origin and not continental European provenance. Native plants have evolved in relative isolation from the Continent so have developed specific characteristics that correspond with the wildlife that depends on them.

Taking the existing conditions in a potential wild-flower garden into account is vital for the garden's success. Take note of the garden's aspect, the soil, areas of light and shade, damp and dry spots and any features such as walls, hedges and trees – and their positioning.

Choose a range of wild-flower species that will produce a floral display throughout the year – some insects begin to forage for nectar in early spring, others continue well into the autumn.

Bees and butterflies appreciate a wide variety of nectar-rich wild flowers rather than highly bred cultivars with double flowers whose complicated structure makes it more difficult for insects to extract the nectar. Sow seed or plant plugs of each type of wild flower in drifts – this makes it easier for bees and butterflies to locate their favourite plants.

Plants that produce abundant seeds attract birds into the garden – teasels, for example, are much loved by goldfinches. Others are worth growing for their foliage – stinging nettles produce leaves that are a rich source of food for caterpillars.

Harebell (*Campanula rotundifolia*, page 193)

Recreating wild-flower habitats in the garden

Meadows
A meadow consists mostly of perennial wild flowers and grasses. This rich variety of flora attracts bees, butterflies including the meadow brown, skipper and marbled white, and other insects that forage on the nectar-rich blooms. In turn, the insects attract small mammals such as hedgehogs

Self-heal (*Prunella vulgaris*, page 182)

and voles, while goldfinches and bullfinches are among the birds that feast on wild-flower seeds.

Recreating a spring or summer hay meadow in the true sense takes several years, but if space allows in the garden it is possible to establish a small area devoted to meadow wild flowers in a relatively short space of time. These species of flowers flourish in full sun and on poor soil. Fertile soil is rich in nitrogen and encourages strong grass growth that eventually overwhelms most wild flowers, although there are wild-flower seed mixes available to suit most soil types. Depending on the choice of wild flowers, seed is sown in spring or autumn – some species benefit from low temperatures before they germinate.

Mowing the meadow area and raking off the hay at the right time of year helps to maintain the diversity of species. If a meadow is neglected it will slowly revert to grass. For a spring meadow, cut and rake in June or July; for a summer meadow cut and rake in August or September. Meadows that include both spring and summer flowers should be cut and raked in August and September. In all cases, the cuts should be delayed until the wild flowers have seeded. Never add fertilisers to meadow wild flowers.

RIGHT Field scabious (*Knautia arvensis*, page 193)

RECOMMENDED PLANTS FOR SPRING WILD-FLOWER MEADOWS

Bugle
(see page 174)

Common cat's ear
(see page 95)

Cowslip
(see page 84)

Daisy
(see page 36)

Lesser stitchwort
(see page 48)

Primrose
(see page 84)

Salad burnet
(see page 164)

Self-heal
(see page 182)

Snake's-head fritillary
(see page 167)

RECOMMENDED PLANTS FOR SUMMER WILD-FLOWER MEADOWS

Common knapweed
(see page 185)

Devil's-bit scabious
(see page 182)

Field scabious
(see page 193)

Goat's-beard
(see page 95)

Greater knapweed
(see page 190)

Harebell
(see page 193)

Lady's bedstraw
(see page 104)

Meadow buttercup
(see page 107)

Musk mallow
(see page 140)

Cornfields

Annual poppies and cornflowers are quintessential wild flowers of cornfields. In a traditional cornfield these annual plants establish themselves on disturbed ground, usually after ploughing. The wildlife that cornfield annuals attract is quite different from that found in meadows. Smaller species of butterflies, hoverflies and bees are drawn by the copious quantities of nectar and pollen, while small mammals such as field mice and voles hide in the dense undergrowth.

Cornfield annuals take only a few months to create a blaze of colour, which will attract a range of insects into the garden almost immediately. They germinate quickly in situ on bare ground or can be grown in containers. Their bright colours and fast establishment make them easy and fun for children to grow. Choose an area of the garden in full sun, ideally where the soil is poor and dry, although cornfield annuals do just as well on fertile soil. Sow seed in spring or autumn but be aware that if sowing in autumn, corn marigold (see list of recommended plants) seedlings will be damaged by frost; poppies, on the other hand, benefit from exposure to cold temperatures. Once flowering is over and the seed has set and dispersed, pull out any dead flower stems and dig over the soil, removing any perennial weeds such as thistles or nettles – this process encourages the annual seeds to germinate the following year. Over time the balance of plants can change as vigorous species, such as corncockle, (see list of recommended plants) start to dominate. Keep these species under control by simply digging up any unwanted seedlings.

RECOMMENDED PLANTS FOR CORNFIELDS

Corn chamomile

Corn marigold (see page 103)

Corncockle (see page 152)

Cornflower

Common field poppy (see page 160)

Lawns

Introducing wild flowers into a lawn enriches what can otherwise be a sterile environment that attracts little wildlife. Infinitely preferable are areas of uncut grass planted with wild flowers that will result in a meadow environment in miniature (see page 240), while even short grass can be improved by planting a variety of low-growing wild flowers that tolerate trampling. However, sowing wild-flower seed on to a nutrient-rich lawn is rarely successful because the grass will eventually overwhelm the wild flowers. Instead plant wild flowers as plugs or transplant young plants grown from seed in seed trays into the lawn. Lawns planted with wild flowers should not be fertilised.

RECOMMENDED WILD FLOWERS FOR SHORT TURF

Bird's-foot trefoil (see page 91)

Chamomile (see page 160)

Daisy (see page 36)

Eyebright (see page 60)

Self-heal (see page 182)

Violets (see pages 167, 170)

Wild thyme (see page 182)

Yarrow (see page 56)

RECOMMENDED WILD FLOWERS FOR LONG GRASS

Devil's-bit scabious (see page 182)

Field scabious (see page 193)

Hogweed (see page 67)

Knapweed species (see pp. 185, 190)

Mallow (see page 140)

Wild teasel (see page 202)

LEFT Eyebright (*Euphrasia* agg., page 60)

Daisy (*Bellis perennis*, page 36)

Ponds

Few natural ponds remain in the wild; most have been filled in, drained or polluted. Consequently the domestic garden pond is one of the most important habitats for protecting species of aquatic and marginal wild flowers, attracting wildlife and providing breeding grounds for amphibians, birds and insects.

Ideally, wildlife ponds should be positioned in full sun with as few overhanging trees as possible and created with at least one muddy border for marginal plants. To ensure a balanced ecosystem, ponds require oxygenators, plants that grow under the surface of the water and supply it with oxygen; floating plants, which have floating leaves and create shade to keep the water cool; emergents, which grow with their roots under the water and their leaves above the surface – these provide landing pads for insects such as dragonflies; and marginals, which grow in the damp soil at the water's edge and form vital habitats and cover for wildlife.

There are plenty of aquatic wild flowers suitable for growing in domestic gardens, but by the same token gardeners must be aware of the handful of exotic plants that are extremely invasive and should be avoided at all costs (see box, right).

PONDS

RECOMMENDED PLANTS: OXYGENATORS

Common water starwort (*Callitriche stagnalis*)

Curled pondweed (*Potamogeton crispus*)

Spiked water milfoil (*Myriophyllum spicatum*)

RECOMMENDED PLANTS: MARGINAL/EMERGENT PLANTS

Brooklime (see page 171)

Bogbean (see page 32)

Cuckoo flower (see page 131)

Flowering rush (see page 140)

Marsh marigold (see page 84)

Meadowsweet (see page 71)

Common water plantain (see page 63)

Yellow flag iris (see page 103)

RECOMMENDED PLANTS: FLOATING-LEAVED PLANTS

Arrowhead (see page 63)

White water lily (see page 67)

INVASIVE PLANTS TO AVOID

Australian swamp stonecrop (*Crassula helmsii*)

Curly waterweed (*Lagarosiphon major*)

Floating pennywort (*Hydrocotyle ranunculoides*)

Green seafingers (*Codium fragile*)

Parrot's feather (*Myriophyllum aquaticum*)

Water fern (*Azolla filiculoides*)

Arrowhead (*Sagittaria sagittifolia*, page 63)

Sweet woodruff (*Galium odoratum*, page 35)

Ramsons or wild garlic (*Allium ursinum*, page 31)

Woodland and shady areas

Woodland wild flowers are some of the most beautiful wild plants. Few gardeners are lucky enough to own a piece of woodland, but with a small shaded area planted with native trees and shrubs, it is possible to nurture a range of shade-loving wild flowers. The key is to establish the correct amount of shade. Most woodland wild flowers bloom in spring when light levels are lower than in summer – if the shade is too dense, then the variety of plants is reduced. The choice of wild flowers determines whether they are planted as bulbs or plants, or sown as seed. Those plants that grow on the edge of a shaded area, where there is more light, are likely to be more successful if sown from seed; in deeper shade, bulbs and young plants establish more easily.

Hedges, walls and fences

Many wild flowers benefit from the shelter
and support provided by a hedge of native
plants, a fence or wall. Boundaries create
nesting sites for birds, bees and small
mammals and protect the garden from winds.
The aspect of the boundary will determine
whether you plant sun-lovers, shade-tolerant
plants or those that thrive in partial shade.
For example, a hedge that runs east to west
can be planted with sun-lovers on the south
side and shade-tolerant plants on the north
side. Some flowers, such as common toadflax
(see page 107) and navelwort (see page 120),
are adept at thriving in wall crevices.

RIGHT Bramble (*Rubus fruticosus*, page 139)

Flower borders

Planting nectar-rich native wild flowers among non-native garden plants encourages bees, butterflies and birds into the garden where they are otherwise put off by the more flamboyant blooms of garden cultivars. Many wild flowers have funnel-shaped flowers, which are especially adapted for long-tongued bees and butterflies. Some, such as the foxglove, are equipped with markings that act like road signs, which guide bees towards the nectar. Plant wild flowers in sunny and sheltered positions in groups rather than in ones or twos so that insects can locate them easily. Choose a range of wild flowers that

RECOMMENDED PLANTS FOR FLOWER BORDERS

Carline thistle (see page 59)

Clustered bellflower (see page 190)

Common comfrey (see page 76)

Field scabious (see page 193)

Foxglove (see page 189)

Greater knapweed (see page 190)

Common mallow (see page 140)

Red clover (see page 160)

Viper's bugloss (see page 176)

RECOMMENDED CULINARY PLANTS

Burnet saxifrage (see page 67)

Stinging nettle (see page 218)

Wild basil (see page 152)

Wild marjoram (see page 152)

Wild strawberry (see page 36)

Wild thyme (see page 182)

produce a floral display throughout the season, supplying nectar for those insects that start to forage in early spring and those that continue late into the autumn.

Culinary and medicinal borders

For centuries, wild flowers have been respected and exploited for their culinary and medicinal uses. Plant a small herb garden with native wild herbs and a selection of plants that can be harvested for culinary use; for the enthusiast a border devoted to medicinal plants could be a novel addition to the garden, and like the wild-flower border described above, both have the added benefit of attracting wildlife into the garden.

RECOMMENDED MEDICINAL PLANTS

Chamomile (see page 160)

Common evening primrose (see page 95)

Feverfew (see page 48)

Ground ivy (cultivar) (see page 170)

Ox-eye daisy (see page 48)

Perennial flax (see page 181)

White deadnettle (see page 32)

White deadnettle (*Lamium album*, page 32)

Where to buy wild flowers and wild-flower seeds

BRITISH WILD FLOWER PLANTS
Burlingham Gardens, 31 Main Road, North Burlingham
Norfolk NR13 4TA Tel 01603 716615 www.wildflowers.co.uk

CHILTERN SEEDS
114 Preston Crowmarsh, Wallingford OX10 6SL
Tel 01491824675 www.chilternseeds.co.uk

NATURESCAPE
Maple Farm, Coach Gap Lane, Langar, Nottinghamshire
NG13 9HP Tel 01949 860592 www.naturescape.co.uk

EMORSGATE WILD SEEDS
Limes Farm, Tilney All Saints, King's Lynn
Norfolk PE34 4RT Tel 01553 829028 www.wildseeds.co.uk

LANDLIFE WILDFLOWERS LTD
Laburnum House, Main Road, Langrick, Boston PE22 7AN
Tel 01205 281902 www.wildflower.co.uk

SHIPTON BULBS
Y Felin, Henllan Amgoed, Whitland, Pembrokeshire SA34 0SL
Tel 01994 240125 www.bluebellbulbs.co.uk

LEFT Carline thistle (*Carlina vulgaris*, page 59)

Useful contacts

**The Botanical Society
of the British Isles**
Botanical Society of
Britain & Ireland
57 Walton Road
Shirehampton
Bristol B211 9TA
Tel 01162 704989
www.bsbi.org

**Conservation Volunteers
Ireland**
Dublin Tel 0749137090
ext. 3602
www.conservationvol-
unteers.ie

English Nature
The Engine House,
Fire Fly Avenue
Swindon
SN2 2EH
Tel 0370 333 1181

Flora Locale
Denford Manor
Hungerford
Berkshire RG7 0UN
Tel 01488 686 186
www.floralocale.org

**Irish Peatland
Conservation Council**
Bog of Allen Nature
Centre
Lullymore
Rathangan
Co. Kildare
Tel (00 353) 045 860133
www.ipcc.ie

Irish Wildlife Trust
Sigmund Business Centre
93A Lagan Road
Dublin Industrial Estate
Glasnevin
Dublin 11
Tel (00 353) 01 8602839
www.iwt.ie

**The National Parks and
Wildlife Service**
7 Ely Place
Dublin 2
Tel (00 353) 1 8883242
www.npws.ie

The National Trust
Heelis
Kemble Drive
Swindon
SN2 2NA
Tel 03448 001895
www.nationaltrust.org.uk

**The National Trust for
Ireland/An Taisce**
Tailor's Hall
Back Lane
Dublin 8
Tel (00 353) 01 454 1786
www.antaisce.org

**The National Trust
Office for Northern
Ireland**
Rowallane Stableyard
Saintfield
Ballynahinch
Co. Down BT24 7LH
Tel 028 9751 0721

**The National Trust
for Scotland**
Hermiston Quay
5 Cultins Road
Edinburgh
EH11 4DF
01314580200
www.nts.org.uk

**The National Trust
Office for Wales**
Priest House
Tredegar House
Newport
South Wales
NP10 8YW
Tel 01633 811659

Plantlife
Brewery House
36 Milford Street
Salisbury
Wiltshire
SP1 2AP
Tel 01722342730
www.plantlife.org.uk

Planta Europa
c/o Plantlife International
Brewery House
36 Milford Street
Salisbury
Wiltshire
SP1 2AP
Tel 01722 342730
www.plantlife.org.uk

**Royal Botanic Gardens
Edinburgh**
20a Inverleith Row
Edinburgh EH3 5LR
Tel 0131 552 7171
www.rbge.org.uk

**Royal Botanic Gardens
Kew**
Richmond
Surrey TW9 3AB
Tel 020 8332 5000
www.rbgkew.org.uk

**Royal Botanic Gardens
Wakehurst Place**
Ardingly
Haywards Heath
West Sussex
RH17 6TN
Tel 01444 894000

**Royal Horticultural
Society**
80 Vincent Square
London SW1P 2PE
Tel 020 3176 5800
www.rhs.org.uk

**Royal Society for the
Protection of Birds**
The Lodge
Sandy
Bedfordshire SG19 2DL
Tel 01767 680551
www.rspb.org.uk

**Trinity College Botanic
Garden**
Botany Department
Trinity College Dublin
Dublin 2
Tel (00 353) 01 8961000
www.tcd.ie

The Wildflower Society
43 Roebuck Road
Rochester
Kent
ME1 1UE
Tel 020 8368 5328
www.thewildflower-
society.com

The Wildlife Trusts
The Kiln
Waterside
Mather Road, Newark
Nottinghamshire
NG24 1WT
Tel 0870 036 7711
www.wildlifetrusts.org

Further reading

BOOKS

A Modern Herbal
Grieve M. (1931, Tiger
Books International)

*Botany for Gardeners:
An Introduction and Guide*
Capon B. (1992, B.T.
Batsford Ltd)

Britain's Rare Plants
Marren P. (1999, T. & A.
D. Poyser)

*British Red Data Books,
1, Vascular Plants.*
Wigginton M.J. (Ed.)
(1999, Joint Nature
Conservation Committee)

*Cassell's Wild Flowers of
Britain & Northern Europe*
Blamey M., Grey-Wilson
C. (2003, Cassell)

*Edible and Medicinal
Plants of Britain and
Northern Europe*
Launert, E. (1981,
Hamlyn)

*Field Guide to the
Wildflowers of Britain.*
Press J.R., Sutton, D.A.,
Tebbs, B.R. (reprint 2002,
The Reader's Digest
Association Ltd).

Flora Britannica
Mabey R. (1997, Chatto
& Windus)

Flora Europaea Vols 1–5.
Tutin, T.G. et al
(1964–80, Cambridge
University Press)

Flora of the British Isles
Clapham, A.R., Tutin,
T.G., Warburg, E.F.
(1962, Cambridge
University Press)

Food for Free
Mabey, R. (1975,
Fontana)

Jekka's Complete Herb Book
McVicar J. (1999, Kyle
Cathie Ltd)

*Kingfisher Field Guide to
Wild Flowers of Britain &
Northern Europe*
Sutton D. (1988,
Kingfisher Books)

*Meadows and Cornfields:
how to create and maintain
a meadow or cornfield to
attract wildlife to your
garden* Steel J. (2001,
Webbs Barn Designs)

*New Atlas of the British
Flora* Preston, C.D.,
Pearman, D.A., Dines,
T.D. (2002, Oxford
University Press)

New Family Herbal
Meyrick, W. (1790,
Thomas Pearson,
Birmingham)

*New Flora of the British
Isles*
Stace, C.A. (Reprinted
2001, Cambridge
University Press)

Plants with a Purpose
Mabey, R. (1979, Fontana)

*Stearn's Dictionary of Plant
Names for Gardeners*
Stearn W. (reprinted
1994, Cassell)

*The Concise British Flora
in Colour* Keble-Martin,
W. (1969, Ebury Press &
Michael Joseph)

The Englishman's Flora
Grigson G. (1955,
Phoenix House Ltd)

The Plant Book.
Mabberley, D.J. (1987,
Cambridge University
Press)

*The RHS A-Z Encyclo-
pedia of Garden Plants*
Brickell C. (ed) (revised
edition 2003, Dorling
Kindersley)

*Wayside and Woodland
Blossoms: a guide to
British wild flowers. Vols I,
II, III* Step E. (1963,
Frederick Warne & Co)

*The Wild Flower Key:
British Isles-N.W Europe*
Rose F. (reissued 1981,
rederick Warne)

*Wildflowers: A descriptive
list of over 140 wildflowers
to attract wildlife to your
garden*
Steel J. (2001, Webbs
Barn Designs)

*The Wildflowers of Britain
and Northern Europe*
Fitter R., Fitter A. (1978,
Collins)

Wildflowers in Colour
Phillips, R. (1977, Pan)

*Wildflowers of Britain
and Europe*
Press, J.R., Gibbons, B.
(1993, New Holland)

*Wildflowers Work: A
technical guide to creating
and managing wildflower
landscapes*
Lickorish S, Luscombe G,
Scott R. (1997, Landlife)

*Wildlife Ponds: how to
create a natural looking
pond to attract wildlife to
your garden*
Steel J.
(2002, Webbs Barn
Designs)

Wild Orchids of Dorset
Jenkinson, M. (1991,
Orchid Sundries Ltd.)

WEBSITES

The Arable Plants Group
www.arableplants.org.uk

BBC
www.bbc.co.uk

Beautiful Britain
www.beautifulbritain.co.uk

**British Society of the
Botanical Isles**
www.bsbi.org.uk

**Convention on
International Trade in
Endangered Species
of Wild Fauna and Flora**
www.cites.org

Defence Estates
www.defence-estates.mod.uk

**Department for
Environment,
Food and Rural Affairs**
www.defra.gov.uk

English Nature
www.english-nature.org.uk

Environment Agency
www.environment-
agency.gov.uk

**Farmed Environment
Company**
www.f-e-c.co.uk

Flora Locale
www.floralocale.org

Garden Links
www.gardenlinks.ndo.co.uk

Habitat
www.habitat.org.uk

**IUCN List of
Threatened Species**
www.redlist.org/

Landlife
www.landlife.org.uk

Naturenet
www.naturenet.net

Naturescape
www.naturescape.co.uk

Natural History Museum
www.nhm.ac.uk

**National Wildflower
Centre**
www.nwc.org.uk

Planta Europa
www.plantaeuropa.org

Plantlife
www.plantlife.org

**Royal Botanic Gardens,
Kew**
www.rbgkew.org.uk

**Royal Horticultural
Society**
www.rhs.org.uk

**Royal Society for the
Protection of Birds**
www.rspb.org.uk

**UK Biodiversity
Action Plan**
www.ukbap.org.uk

The Wildlife Trusts
www.wildaboutgardens.org
www.wildflowerlinks.co.uk

The Wildlife Trusts
www.wildlifetrust.org

Wriggly Wrigglers
www.wrigglywigglers.co.uk

Index of plants by common name

Index of plants by Latin name

A

Aceras anthropophorum 213
Achillea ptarmica 71
Aconitum napellus 194
Adoxa moschatellina 210
Aegopodium podograria 47
Aethusa cynapium 32, 68
Agrimonia eupatoria 103
Agrimonia procera 108
Agrostemma githago 13, 21, 152, 231,
    239, 241
Ajuga reptans 16, 174, 240, 245
Alchemilla alpina 107
Alchemilla mollis 218, 221
Alchemilla vulgaris 107, 221
Alchemilla xanthochlora 218, 221
Alisma plantago-aquatica 63, 243
Alliaria petiolata 19, 31
Allium schoenoprasum 186
Allium scorodoprasum 186
Allium triquetrum 32
Allium ursinum 31, 244, 245
Althaea officinalis 156
Anacamptis pyramidalis 132
Anagallis arvensis 164
Anagallis tenella 144
Anchusa arvensis 178
Anemone nemorosa 16, 30, 245
Angelica archangelica 68
Angelica sylvestris 68
Antennaria dioica 39
Anthemis arvensis 13, 21, 152, 231, 239,
    241
Anthemis cotula 60
Anthriscus caucalis 32
Anthriscus sylvestris 19, 32, 47, 52, 147
Anthyllis vulneraria 94, 235
Antirrhinum majus 159
Aphanes arvensis 226
Apium graveolens 55, 71
Apium nodiflorum 71
Aquilegia vulgaris 173
Arabidopsis thaliana 39
Arctium lappa 197
Arctium minus 197
Arctium nemorosum 197
Arctostaphylos uva-ursi 128
Arenaria serpyllifolia 59
Armeria maritima 127
Armoracia rusticana 36
Artemisia absinthium 124
Artemisia vulgaris 111
Arum italicum 209
Asparagus officinalis subsp. prostratus 210
Asperula cynanchica 147
Aster novi-belgii 193, 206
Aster tripolium 193, 206
Astragalus glycyphyllos 197, 229
Atriplex laciniata 225
Atriplex littoralis 228
Atriplex prostrata 225
Atropa belladonna 172, 217
Azolla filiculoides 243

B

Baldellia ranunculoides 131
Ballota nigra 197
Bellis perennis 15, 16, 36, 240, 242, 243
Berula erecta 59, 71
Beta vulgaris subsp. maritima 214
Blackstonia perfoliata 104
Borago officinalis 181
Brassica napus 83
Brassica napus subsp. oleifera 83
Brassica oleracea
Brassica rapa 111
Bryonia dioica 56
Buddleja davidii 202
Buglossoides purpurocaerulea 174
Butomus umbellatus 140, 240

C

Cakile edentula 151
Cakile maritima 151
Calamintha ascendens 144
Calla palustris 68
Callitriche stagnalis 243
Calluna vulgaris 198
Caltha palustris 15, 84, 243
Calystegia sepium 64
Calystegia soldanella 132
Campanula glomerata 190, 235, 247
Campanula latifolia 186
Campanula ranunculoides 185
Campanula rotundifolia 189, 193, 240
Campanula trachelium 186
Cannabis 63, 143
Capsella bursa-pastoris 76
Cardamine amara 36
Cardamine flexuosa 60
Cardamine hirsuta 60
Cardamine pratensis 131, 243
Carduus crispus 186
Carduus nutans 189
Carlina vulgaris 59, 247
Centaurea 242
Centaurea nigra 185, 240
Centaurea scabiosa 190, 214, 240, 247
Centaurium erythrea 139
Centranthus ruber 148
Cephalanthera damasonium 76
Cerastium arvense 44
Cerastium fontanum 44
Chaenorhinum minus 205
Chaerophyllum temulum 52, 147
Chamaemelum nobile 60, 242, 247
Chamerion angustifolium 21, 139
Chelidonium majus 91
Chenopodium album 13, 217
Chenopodium bonus-henricus 221
Chrysanthemum segetum 13, 21, 103,
    239, 241
Chrysosplenium alternifolium 80
Chrysosplenium oppositifolium 80
Cichorium intybus 190, 239
Circaea lutetiana 52, 245

C (continued)

Cirsium acaule 205
Cirsium arvense 202
Cirsium helenioides 185
Cirsium palustre 8, 186
Cirsium vulgare 197, 202
Claytonia perfoliata 31
Claytonia sibirica 128
Clematis vitalba 19, 79, 245
Clinopodium vulgare 152, 247
Cochlearia anglica 35
Cochlearia danica 35
Cochlearia officinalis 36
Codium fragile 243
Coeloglossum viride 213, 236
Colchicum autumnale 155
Conium maculatum 32, 59
Conopodium majus 40
Convallaria majalis 39
Convolvulus arvensis 21, 64, 132
Corydalis lutea 92
Coryza canadensis 60
Crambe maritima 14, 19, 51
Crassula helmsii 231, 243
Crepis capillaris 108
Crithmum maritimum 14, 15, 104, 112
Crocosmia aurea 163
Crocosmia pottsii 163
Crocosmia x crocosmiflora 163
Crocus nudiflorus 155
Crocus sativus 155
Cruciata laevipes 94
Cuscuta epithymum 147, 231
Cymbalaria muralis 171
Cynoglossum officinale
Cypripedium calceolus 232

D

Dactylorhiza cruenta 132
Dactylorhiza fuchsii 131, 178
Dactylorhiza incarnata 132
Dactylorhiza praetermissa 178
Daucus carota 55
Dianthus armeria 139, 231
Dianthus deltoides 135
Dianthus gratianopolitanus 156
Digitalis purpurea 189, 247
Diplotaxis muralis 91, 100
Diplotaxis tenuifolia 91
Dipsacus fullonum 75, 202, 239, 240, 242
Dipsacus pilosus 75
Doronicum pardialanches 84
Drosera 19, 124
Drosera intermedia 72
Drosera rotundifolia 72
Dryas octopetala 76

E

Echium vulgare 19, 176–177, 247
Epilobium hirsutum 139
Epipactis helleborine 229

E (continued)

Epipactis palustris 68
Eranthis cilicica 80
Eranthis hyemalis 80
Erica cinerea 206
Erica tetralix 155
Erigeron acer 181
Erigeron borealis 205
Erigeron karvinskianus 44
Erodium cicutarium 155
Erophila verna 35
Eryngium maritimum 19, 201
Erysimum cheiri 84
Eupatorium cannabinum 21, 68, 143
Euphorbia exigua 222
Euphorbia helioscopia 222
Euphorbia lathyris 221
Euphorbia paralias 222
Euphorbia peplus 221
Euphorbia portlandica 222
Euphrasia agg. 14, 60, 242

F

Fallopia convolvulus 217
Fallopia japonica 72
Filago vulgaris 104
Filipendula ulmaria 52, 71, 243
Filipendula vulgaris 52
Foeniculum vulgare 123
Fragaria vesca 16, 36, 247
Fritillaria meleagris 16, 167, 236, 240
Fumaria officinalis 148

G

Galanthus elwesii 31
Galanthus nivalis 16, 31, 232, 245
Galanthus plicatus 31
Galega officinalis 194
Galeopsis tetrahit 63
Galinsoga parviflora 71
Galinsoga quadriradiata 71
Galium aparine 56, 64
Galium boreale 42
Galium mollugo 48
Galium odoratum 35, 244, 245
Galium palustre 56
Galium saxatile 64
Galium uliginosum 56, 60
Galium verum 16, 104, 240
Gentiana nivalis 205
Gentiana pneumonanthe 198
Gentiana verna 174
Gentianella amarella 206
Geranium 13
Geranium dissectum 144
Geranium endressii 44
Geranium lucidum 128
Geranium molle 147
Geranium pratense 190
Geranium pyrenaicum 193
Geranium robertium 148

Acknowledgements

My thanks and appreciation to my mother, Jacquie, who taught me to love plants and words; to bring them together in this book would never have been possible without her.

To Sarah Cuttle for her hard work, patience and for making this book so beautiful with her incredible photography.

To Barry Tebbs, botanical advisor and mentor throughout the long journey to produce this book – for his encouragement, moral support, advice and vital sense of humour, and for sharing, so generously and unconditionally, his immense knowledge with me.

To Kyle Cathie and Muna Reyal for their infinite patience and understanding when the going got tough.

To all those that have gone before me and written the many works I used as reference and from which I learned so much about these precious wild flowers and their habitats, not least to respect and appreciate them for the priceless part they play in making our natural world go round.

And finally to Simon and Samuel, my boys, for their love, support and for making it all worthwhile.

Rae Spencer-Jones

There are many people I want to thank in the production of this book, starting with Kyle Cathie herself for putting the idea in our heads in the first place. Also to Muna Reyal for her marvellous efficiency, calm and good humour throughout.

An especially big thank you to my wonderful sister Lou, who worked with me on a great deal of the photography, sharing the highs and tolerating me through the lows. She navigated me over a great many miles, often keeping a beady eye out on the passing hedgerows, for various absentee plants.

Also to Linda Laxton at British Wildflower Plants. I can truly say that without Linda's in-depth knowledge and endless optimism, I would have faced an impossible uphill struggle. Her family, too, were wonderful in coping without her on the afternoons that we spent hunting around one of the local conservation areas.

It was also a great pleasure and privilege to accompany Dr Tom Cope and Barry Tebbs on some of their many wild-flower hunts across the country. Not only have they both given freely of their time and expertise, but also very generously let me raid their personal library collection, to fill in where I failed in my task.

My gratitude is also due to all the other botanists and nurseries up and down the country who helped me in my quest, especially to Dr David Shaw in Wales, for skilfully guiding myself and Linda up a mountain in Snowdonia in the rain in search of a water lobelia.

All in all, this has been an intriguing project, and not least because of the wonderful people I have been lucky enough to work with. I am now no longer able to pass a patch of land, be it in the country or the city, without having a close look at what wild flower is making itself at home.

Sarah Cuttle

Picture credits

P32 pic (6) T.A. Cope; 50 (1) T.A. Cope; 58 (4) T.A. Cope; 58 (6) T.A. Cope; 65 (6) T.A. Cope; 73 (2) Barry Tebbs; 73 (3) T.A. Cope; 77 (1) T.A. Cope; 77 (6) T.A. Cope; 79 (3) John Feltwell/Garden Matters; 81 (1) T.A. Cope; 113 (2) T.A. Cope; 118 (5) Barry Tebbs; 121 (1) Heather Angel/Natural Visions; 121 (2) T.A. Cope; 125 (3) T.A. Cope; 125 (4) Howard Rice/The Garden Picture Library; 129 (1) T.A. Cope; 133 (3) T.A. Cope; 141 (4) T.A. Cope; 157 (5) David Askham/The Garden Picture Library; 169 (3) T.A. Cope; 188 (3) T.A. Cope; 188 (5) T.A. Cope; 195 (4) T.A. Cope; 200 (5) T.A. Cope; 204 (1) T.A. Cope; 204 (2) T.A. Cope; 204 (3) Barry Tebbs; 212 (4) T.A. Cope; 215 (2) Barry Tebbs; 219 (6) T.A. Cope; 223 (2) T.A. Cope; 223 (5) T.A. Cope; 223 (6) T.A. Cope; 227 (4) T.A. Cope.